# Innovations of Green Process Engineering for Sustainable Energy and Environment 2013

**Topical Conference at the 2013 AIChE Annual Meeting: Global Challenges for Engineering a Sustainable Future**

## San Francisco, California, USA
## 3-8 November 2013

ISBN: 978-1-63439-032-3

**Printed from e-media with permission by:**

Curran Associates, Inc.
57 Morehouse Lane
Red Hook, NY 12571

**Some format issues inherent in the e-media version may also appear in this print version.**



Printed by Curran Associates, Inc. (2014)



AIChE
3 Park Avenue
New York, NY 10016-5991

Phone: (203) 702-7660
Fax: (203) 775-5177

www.aiche.org

**Additional copies of this publication are available from:**

Curran Associates, Inc.
57 Morehouse Lane
Red Hook, NY 12571 USA
Phone: 845-758-0400
Fax: 845-758-2634
Email: curran@proceedings.com
Web: www.proceedings.com

# TABLE OF CONTENTS

**(32a) Overview Of Materials In The Built Environment** ... 1
*Ron Judkoff*

**(32b) Energyplus Model-Based Predictive Control (EPMPC) Using Matlab/Simulink and Mle+** ... 2
*Jie Zhao, Khee Poh Lam, B. Erik Ydstie*

**(32c) The Sol-Char Toilet: Concentrated Solar Power Transported By Fiber Optics for Pyrolysis of Human Waste to Biochar** ... 3
*Richard P. Fisher, Allan Lewandowski, Tesfayohanes W. Yacob, Ryan B. Mahoney, Dragan Mejic, Cori Oversby, Sara Beck, Barbara J. Ward, Josh Kearns, R. Scott Summers, Karl G. Linden, Alan W. Weimer*

**(32d) Building-Integrated Active Modular Phytoremediation System** ... 4
*Anna Dyson, Ahu Aydogan Akseli, Jason Vollen, Marianne Nyman, Jeffrey Bird*

**(32e) Integrating Technological and Ecological Design for Sustainable Buildings On a University Campus** ... 5
*Erin L. Gibbemeyer, Bhavik R. Bakshi*

**(32f) Energy Conservation Approach to Cooling Process By Employing Preheat Tank and Chilled Water Tanks** ... 6
*Dr. Robert W. Peters, Zhuo Li*

**(32g) Operation of Residential Hybrid Renewable Energy Systems: Integrating Forecasting, Optimization and Demand Response** ... 7
*Xiaonan Wang, Nael H. El-Farra, Ahmet Palazoglu*

**(36a) Green Chemistry, Green Engineering, Green Economy At US EPA 's Sustainable Technologies Division** ... 9
*John Leazer*

**(36b) Sustainability Practice in Industry** ... 10
*Joseph B. Powell*

**(36c) Green Engineering and Sustainability in the Chemical Engineering Curriculum** ... 11
*David R. Shonnard, David T. Allen, Nhan Nguyen, Sharon Austin*

**(36d) Driving Private Sector Adoption of Greener Processes and Technologies** ... 12
*Abby Lindsay*

**(36e) Minimization of the GHG Emissions At the Macroeconomic Level Via a Multi-Objective Input-Output Approach: A Case Study of the EU-25 Economy** ... 13
*Daniel Cortés, Antonio Ruiz-Hernández, Gonzalo Guillén-Gosálbez, Maria Llop, Roger Guimerá, Marta Sales-Pardo*

**(36f) Economic Analysis and Optimization of Carbon Capture Storage Pipeline Transport Network System** ... 15
*Seungkyu Dan, Byung Joon Kang, Seonghyun Cho, En Sup Yoon*

**(36g) Economic Development Through Entrepreneurship and Technology Innovation: Bringing Together Industry, Academia, and Government** ... 16
*Nada Marie Anid*

**(37a) The Carbon Capture Simulation Initiative (CCSI): Computational Tools To Accelerate Commercial Development** ... 17
*David C. Miller*

**(37b) The Four Dimensional Product: Integration Over Time Is the Only Way to Understand Sustainability** ... 18
*William Banholzer, Mark E. Jones*

**(37c) Maximization of Energy Efficiency of a Combined Heat and Power Plant** ... 19
*Thomas F. Edgar, Kody Powell, Jong Kim, Kriti Kapoor*

**(37d) Pan American Issues on Biofuel Sustainability** ... 20
*David R. Shonnard*

**(50a) Crosslinked Anion Exchange Membranes for Redox Flow Batteries** ... 21
*Min-suk Jung, Janvier Parrondo, Christopher Arges, Vijay Ramani*

**(50b) Challenges to Implementation of High-Rate Energy Storage in Military Applications** ... 22
*John Heinzel, Jason Ostanek, Donald Hoffman, Ian Peek*

**(50c) Novel Pt/TixM1-xO2 Nanocatalysts for Fuel Cells** ... 23
*Bing Joe Hwang, Van Thi Thanh Ho, Chun-Jern Pan, Wei-Nien Su*

**(50d) A Tunable Hierarchical Porous Carbon Structure for Studying Electrochemical Energy Conversion** ... 24
*Kwong-Yu Chan, Chunzhen Yang*

**(50e) High-Performance Electrodes for Hydroxide Exchange Membrane Fuel Cells** .......... 26
*Robert B. Kaspar, Michael Letterio, Kurt Jensen, Shuang Gu, Yushan Yan*
**(50f) Capacitive Mixing in Porous Electrodes for Energy Production** .......... 27
*J.F. Gabitto, C. Tsouris*
**(74a) Process Systems Engineering Of Ionic Liquids** .......... 28
*Suojiang Zhang, Xiangping Zhang, Ying Huang, Yajing Xu*
**(74b) The Word Green, What Does It Mean?** .......... 31
*Dickson E. Ozokwelu*
**(74c) Energy, Economy, Environment and Sustainable Development in the Middle East and North Africa** .......... 32
*M. R. Riazi*
**(74d) Advanced Zeolite Membrane-Enabled Air Dehumidification and Conditioning Technologies for Buildings** .......... 33
*Wei LIU*
**(74e) Ionic Liquids In Biorefining and CO2 Capture** .......... 34
*Jyri-Pekka Mikkola*
**(74f) Process Intensification and Coupling Effects for Efficient Conversion of Biomass Based On Ionic Liquids** .......... 35
*Xuehui Li, Jinxing Long, Hao Ma*
**(118a) Sustainability Progress** .......... 36
*Martin A. Abraham*
**(118b) Sustainable Biofuels: A Case Of Ethics, Politics, Economics and (I almost forgot) Technology** .......... 37
*John J. Sheehan*
**(118c) NSF and Support Of Sustainable Engineering Research and Education** .......... 38
*Bruce Hamilton*
**(118d) Sustainable Development In The Face Of Water Stress** .......... 39
*Danny D. Reible*
**(146a) Development of Composite Hollow Fiber Membranes for Aqueous and Non-Aqueous Nanofiltration** .......... 40
*Shi-Peng Sun, Sui Yung Chan, Tai Shung Chung*
**(146b) Self-Heat Recuperative Seawater Desalination Process Using Fluidized Bed** .......... 41
*Hiroyuki Mizuno, Yasuki Kansha, Masanori Ishizuka, Atsushi Tsutsumi*
**(146c) Characteristics of Water and Salt Transport, Free Volume and Their Relationship With the Functional Groups of Novel Cellulose Esters** .......... 42
*Rui Chin Ong, Bradley J. Helmer, Jos S. de Wit, Tai Shung Chung*
**(146d) Bioremediation of Cu(II) Contaminated Aqueous Solution Using Acinetobacter Sp** .......... 43
*Subhajit Majumder, Gayathri Gangadhar, Smita Raghuvanshi, Suresh Gupta*
**(146e) Calcium Ions in Gypsum (CaSO4 2H2O) Scaling in Forward Osmosis Processes and a Novel Design Strategy to Fabricate Polybenzimidazole Anti-Scaling Hollow Fiber Membranes** .......... 44
*Si Cong Chen, Jincai Su, Feng-Jiang Fu, Baoxia Mi, Tai-Shung Chung*
**(146f) New Insights Into Hydrophobic/Hydrophilic Composite Hollow Fibers for Direct Contact Membrane Distillation** .......... 46
*Jia Ming Zhu, Lan Ying Jiang*
**(182a) Economic and Engineering Assessment of Combined Heat and Power Fuel Cell Systems Installed in Commercial Buildings** .......... 47
*Kriston Brooks, Siva Pilli*
**(182b) Systematic Approach for Energy Efficient Eco-Industrial Parks** .......... 48
*Mahmoud Bahy Noureldin, Saleh A Qahtani, Mana Owaidah, Abdulaziz Nutaifi*
**(182c) The Potential Of Osmotic Membrane Dehumidification** .......... 50
*Arthur S. Kesten, Jeffrey R. McCutcheon, Ariel Girelli, Jack N. Blechner*
**(182d) Improvement of Distillation Column Efficiency By Integration With Organic Rankine Power Generation Cycle** .......... 52
*Dmitriy A. Sladkovskiy*
**(182e) High Surface Area Zeolite Membrane for Efficient Bio-Fuel Production** .......... 53
*Wei LIU*
**(182f) Syngas Mass Transfer Analysis in a Hollow Fiber Reactor** .......... 54
*James J. Orgill, Randy S. Lewis, Hasan K. Atiyeh*
**(182g) An Innovative Methanol Synthesis Process Using Self-Heat Recuperation** .......... 55
*Yasuki Kansha, Masanori Ishizuka, Atsushi Tsutsumi*
**(253a) Molecular Simulations Of Ionic Liquids** .......... 56
*Suojiang Zhang, Kun Dong*

**(253b) Monitoring Inter- and Intra-Molecular Interactions in Imidazolium Protic Ionic Liquids HCnImNTf2 (n=0-12)** ........ 57
*Vladimiros Nikolakis, Anastasia Maria Moschovi, Vassilios Dracopoulos*

**(253c) Beyond 1,3-Difunctionalized Imidazolium Cations: Properties of Ionic Liquids With Tri-, Tetra- and Pentafunctionalized Cations** ........ 58
*Jason E. Bara, Matthew S. Shannon*

**(253d) Volumetric Properties of Mixtures of Ionic Liquids With CO2 Or CO2 + Ethanol** ........ 59
*Heather Grandelli, Erdogan Kiran*

**(253e) Molar Volume Dependence of N2 Solubility in Ionic Liquids** ........ 60
*Joshua Bennett, Aruni DeSilva, Joan F. Brennecke*

**(253f) Structure and Properties of Ionic Liquid/Organic Co-Solvent Mixtures** ........ 61
*Santosh Bandlamudi, Michael Cooney, Peter Chupas, Kenneth M. Benjamin*

**(253g) The Interactions of the Corn Protein Zein and Ionic Liquids** ........ 62
*Sean Tomlinson, John R. Schlup, Jennifer L. Anthony*

**(253h) Synthesis and Characterization of Novel Ternary Deep Eutectic Solvents** ........ 63
*Yuting Liu, Yanjun Xing*

**(311b) Systematic Molecular Simulation Investigation of the Influence of the Chemical Structure of Anion and Cation to the Properties of Imidazolium-Based Ionic Liquids and to the Thermodynamic Properties of Light Gases Dissolved Therein** ........ 64
*Ioannis G. Economou, Eleni Androulaki, Niki Vergadou*

**(311c) Distinctive Nanoscale Organization of Double- Cationic Versus Single-Cationic Ionic Liquids** ........ 65
*Song Li, Guang Feng, Jose Leo Banuelos, Gernot Rother, Pasquale F. Fulvio, Sheng Dai, Peter T. Cummings*

**(311d) Aggregation in Ionic Liquids: Molecular Dynamics Simulation** ........ 66
*Xiaomin Liu, Xiaochun Zhang, Suojiang Zhang*

**(311e) Modeling Chemical and Physical Interactions in CO2/Ionic Liquids With the Reaxff Force Field** ........ 67
*Bo Zhang, Adri C.T. van Duin, J. Karl Johnson*

**(311f) Prediction of Thermodynamic Properties of Ionic Liquids: Improved Performance From Ab Initio Molecular Dynamics** ........ 68
*Yong Zhang, Edward J. Maginn*

**(311g) Molecular Thermodynamic Characterization of Ionic Liquids for Separation Processes** ........ 69
*Lourdes F. Vega, Felix Llovell, Mariana B. Oliveira, Oriol Vilaseca, Rosa M. Marcos, Joao A.P. Coutinho*

**(367a) Phase Behavior of Gases in Ionic Liquids** ........ 71
*Mark B. Shiflett, Akimichi Yokozeki*

**(367b) The Viscosity of Ether-Functionalized Ionic Liquids** ........ 72
*Zhengjian Chen, Tong Xue, Jong-Min Lee*

**(367d) Time-Dependent Diffusion and Chemical Exchange in Mixtures of An Amine-Functionalized Ionic Liquid and Carbon Dioxide By NMR** ........ 73
*Eric Hazelbaker, Han Wang, Samir Budhathoki, Jindal K. Shah, Edward J. Maginn, Sergey Vasenkov*

**(367e) Investigation of the Feasibility of Flue Gas Separations Using An Ionic Liquid From a Thermodynamic Point of View** ........ 74
*Sona Raeissi, Louw J. Florusse, Cor J. Peters*

**(367f) Free Volume Basis of Gas Solubility and Selectivity in Functionalized Imidazolium-Based Ionic Liquids** ........ 76
*Matthew S. Shannon, Jason E. Bara, W. Jeffrey Horne*

**(367g) The Effects of Water Concentration On the Properties of Imidazolium-Based Ionic Liquids** ........ 77
*Amir Abolhasani Niazi, Andrew Crothers, Brooks D. Rabideau, Ahmed E. Ismail*

**(387a) Findings On Life Cycle Aquatic Ecotoxicity Impacts of Ionic Liquids** ........ 78
*Amirhossein Mehrkesh, Arunprakash T. Karunanithi*

**(387b) Comparative Life Cycle Assessment of Vaccine Cold Chain Infrastructure in Developing Countries** ........ 79
*Bahador Mousavi, Arunprakash T. Karunanithi*

**(387c) Hybrid Life Cycle Assessment Model of Silicon Photovoltaics** ........ 80
*Yuan Yao, Yuan Chang, Eric Masanet*

**(387d) Root-Cause Analysis in the Design of Chemical Processes** ........ 82
*Preeti Gangadharan, Liwen Chen, Helen H. Lou*

**(387e) Development of Product-Based Energy Intensity Metrics for the California Food Industry** ........ 83
*Michael E. Walker, Margot Hutchins, Dave Lettieri, Vince Romanin, Paul Blinde, Eric Masanet*

**(387f) Estimations of Direct and Indirect Emissions From Mobile Air Conditioner Units in Automobile Transportation: The Cool Mile Functional Unit** ........ 84
*Paul Blowers, Ibrahim Alayedi, Austin Dougless, Brian Gerwe, Adam Mesta, Caleb Rugel, Yi Yang Wang*

**(387g) Regional Air Quality Improvement By Scheduling of Multiple Chemical Plant Start-Ups** .......... 85
*Tianxing Cai, Qiang Xu*
**(437a) Gas Purification and Separation With Ionic Liquids** .......... 86
*Xiangping Zhang, Haifeng Dong, Ying Huang, Suojiang Zhang, Xin Zhang*
**(437b) Are Ionic Liquids "Drop-In" Replacements for Conventional CO2 Capture Solvents in Packed Absorption Towers? Capital Considerations** .......... 88
*David L. Minnick, Lawrence R. Weatherley, Aaron M. Scurto*
**(437c) Post-Combustion CO2 Capture With Ionic Liquids Involving Phase Change** .......... 89
*Samuel Seo, Luke D. Simoni, Aruni DeSilva, Marjorie Massel, Mark A. Stadtherr, Joan F. Brennecke*
**(437d) Bubble Behavior in Seperation Process With Ionic Liquids** .......... 90
*Haifeng Dong, Xin Zhang, Di Bao, Xiangping Zhang, Suojiang Zhang*
**(437e) Permeability of Carbon Dioxide Through Ionic Liquid Membrane With Polyvinylidene Fluoride** .......... 91
*Yusuke Shimoyama, Shogo Komuro, Pinyarat Jindaratsamee*
**(437f) Phase Equilibria and Separations Involved in the Synthesis of the Ionic Liquid 1-Hexyl-3-Methylimidazolium Bromide ([HMIm][Br]) in Acetone** .......... 92
*David L. Minnick, Aaron M. Scurto*
**(437g) A Promising Electrolyte of Ionic Liquids for Aluminium Deposition** .......... 93
*Qian Wang*
**(444a) Novel Catalysts and Reactor Concepts for Sustainable Processes: the Use of LCA Data in Process Design** .......... 95
*Polina Yaseneva, Alexei Lapkin*
**(444b) Techno-Economic and Environmental Assessment of Ionic Liquid-Based Biorefineries** .......... 98
*Sara Giarola, Nilay Shah*
**(444c) Optimal Simultaneous Production of Bio-Ibutene and Bioethanol From Switchgrass** .......... 100
*Mariano Martín, Ignacio E. Grossmann*
**(444d) Synthesis and Design of Biorefinery Processing Networks With Uncertainty and Sustainability Analysis** .......... 102
*Peam Cheali, Krist V. Gernaey, Gürkan Sin*
**(444e) Ecosystems As Unit Operations for Innovative and Sustainable Process Design** .......... 104
*Bhavik R. Bakshi, Robert A. Urban, Guy Ziv*
**(444f) Process Analysis of Continuous Catalytic Gasification As a Waste to Energy Alternative** .......... 105
*Charles F. Tillie, Melissa J. Riedthaler, Shreya Adhikari, Stephen A. Reeves, Jorge E. Gatica*
**(445a) Ethyl Acetate Synthesis Using Semi-Batch Reactive Distillation** .......... 106
*Yu-Lung Kao, Li-Wen Hsu, Jeffrey D. Ward*
**(445b) Coupled Permsmbr - Process Design and Development for 1,1-Dibutoxyethane Production** .......... 107
*Carla S. Pereira, Viviana T. Silva, Alirio E. Rodrigues*
**(445c) 2-Acetoxy Propionic Acid Using Reactive Distillation – An Intermediate in Synthesis of Bio-Acrylic Acid From Lactic Acid** .......... 108
*Aspi K. Kolah, Carl T. Lira, Dennis J. Miller*
**(445d) Novel Technology for Producing High Purity n-Hexane and Benzene** .......... 109
*Yingzhe Yu, Yonghui Li, Zhongfeng Geng, Jing Ma, Zhiqiang Zhang, Minhua Zhang*
**(445e) Online Monitoring of Emulsion Polymerisations Using Raman Spectroscopy** .......... 110
*Claudia Houben, Gabit Nurumbetov, David Haddleton, Alexei Lapkin*
**(445f) Process Intensification in Millistructured Extraction Columns** .......... 111
*Alexander Holbach, Erkan Caliskan, Norbert Kockmann*
**(445g) A NOVEL Comparison Among Intensified and Traditional Complete Ethyl Acetate Productions Designs** .......... 113
*Miguel Santaella, Paulo C. Narvaez, Alvaro Orjuela*
**(478a) Enhanced Yield of Bio-Gas Production With Modified Themo-Reactor Configuration and Separation of CO2** .......... 114
*Muhammad suleman Tahir, Mahmood Saleem*
**(478b) Microwave-Assisted Catalytic Glycolysis of Poly(ethylene terephthalate) By Using Ionic Liquids** .......... 115
*Yuting Liu, Yanjun Xing*
**(478c) Degradation of Rhodamine-B By Heteropolyoxometalate Ionic Liquid Under Visible Light** .......... 116
*Yuting Liu, Yanan Chen, Yanjun Xing*
**(478d) Metal-Containing Ionic Liquid As An Effective Catalyst for the Degradation of Poly(ethylene terephthalate) in Ethylene Glycol** .......... 117
*Xingmei Lu, Qian Wang, Suojiang Zhang*
**(478e) Anaerobic Digestion Technology:The Energy,Environment and Health of Developing Nations** .......... 118
*Oluyemi Adetule*

**(478f) Phosphorus Recovery From Municipal Wastewater Treatment Sludge** .......... 119
*Kamal Lamichhane Upadhyaya, W. Todd French, Rafael Hernandez*
**(478g) Techno- Economic Analysis of Wastewater Biosolids Gasification** .......... 120
*Nicholas P.G. Lumley, Jason M. Porter*
**(501a) Low-Temperature and Low-Pressure Fuel Hydrodesulfurization By Solid Catalyst Coupling With Ionic Liquids** .......... 121
*Chunshan Li, Suojiang Zhang*
**(501b) Reactivity of Biopolymers in Ionic Liquids** .......... 122
*Jason P. Hallett, Gilbert de Gregorio, Nur Liyana Ismail*
**(501c) Microwave-Assisted Ionic Liquid Pretreatment and Catalysis of Lignin to Phenol in a Biphasic Reaction System** .......... 123
*Jian Xu*
**(501d) Short- and Long-Term Thermal Stabilities Of Imidazolium-Type Ionic Liquids On Catalytic Metal-Oxide Supports** .......... 124
*Asli Akcay, Volkan Balci, Alper Uzun*
**(501e) Cost-Efficient Ionic Liquid and Water Mixtures for Biomass Dissolution** .......... 126
*Anthe George, Kim Tran, Noppadon Sathitsuksanoh, Ning Sun, Niall Mac Dowell, Seema Singh, Blake Simmons, Tom Welton, Jason P. Hallett*
**(501f) Conversion of Residual Ethylene Oxide in the Ionic Liquid Catalyzed CO2 Cycloaddition Process** .......... 127
*Jian Sun, Weiguo Cheng, Jinquan Wang, Junping Zhang, Suojiang Zhang*
**(501g) Aggregation and Solubilization Phenomena During Ionic Liquid Co-Solvent Based Lipid and Phorbol Ester Extraction From Biomass** .......... 128
*Santosh Bandlamudi, Michael Cooney, Kenneth M. Benjamin*
**(519b) Biorefinery: Liquefaction of Pyrolysis Char to Biofuel** .......... 129
*Roland Feiner, Nikolaus Schwaiger, Hannes Pucher, Peter Pucher, Matthäus Siebenhofer*
**(519c) Mechanistic Insight Into Sonochemical Biodiesel Synthesis Using Heterogeneous Base Catalyst** .......... 131
*Hanif A Choudhury, Vijay Moholkar*
**(519d) Energetic and Environmental Assessment of Biomass Processing in the Sugar-Cane Industry** .......... 132
*Michel Kahwaji Janho, Quinton A. Shuster, Jorge E. Gatica, María Rosa Hernández, Fernando Daniel Mele, Mauricio Colombo*
**(519e) Investigation of Solvent System for the Production of Biodiesel From Sludge Palm Oil (SPO) By Enzymatic Transesterification** .......... 133
*Md. Zahangir Alam, Ricca Rahman Nasaruddin, Mohammed Saedi Jami, Mohammad Shahab Uddin*
**(519f) Update to Biodiesel Production From Activated Sludge and Economic Analysis** .......... 134
*Adebola T. Coker, Rafael Hernandez, Emmanuel D. Revellame, William E. Holmes, W. Todd French*
**(519g) An Intensified Integrated Biodiesel Manufacturing Process for Conversion of Waste Oils and Alcohols Into Biodiesel to Power Campus Bus System** .......... 135
*Aydin K Sunol, Zachary Cerniga, Eilis Mcgranaghan*
**(534a) Plate Food Waste: A Valuable Source of Energy** .......... 136
*Stephanie Jung*
**(534b) Biofuel Production and Kinetics Study of Catalytic Microwave Pyrolysis of Douglas Fir Pellet Over Activated Carbon Supported Metal Catalyst** .......... 137
*Quan Bu, Hanwu Lei, Lu Wang, Juming Tang*
**(534c) Molecular-Level Modeling of Municipal Solid Waste Gasification** .......... 138
*Scott R. Horton, Yu Zhang, Craig A. Bennett, Michael T. Klein, Frank Petrocelli*
**(534d) Performance Analysis of Rdf Gasification in a Two Stage Fluid Bed - Plasma Process** .......... 140
*Massimiliano Materazzi, Paola Lettieri, Chris Chapman, Richard Taylor*
**(534e) Process Modeling of Continuous Catalytic Gasification As a Waste to Energy Alternative** .......... 143
*Charles F. Tillie, Melissa J. Riedthaler, Stephen A. Reeves, Shreya Adhikari, Jorge E. Gatica*
**(534f) Performance Evaluation of Accelerated Carbonation of Basic Oxygen Furnace Slag Via a Rotating Packed Bed: Modeling, Analysis and Maximization** .......... 144
*Shu-Yuan Pan, Pen-Chi Chiang, Yi-Hung Chen, E-E Chang, Chung-Sung Tan*
**(534g) Production of Alumina From Fly Ash By a Mild and Cleaner Hydrometallurgical Process** .......... 145
*Li Zhong, Dongping Duan, E Zhou, Hongliang Han*
**(564a) Strategies in the Sustainable Design of Active Pharmaceutical Ingredients for Solvent Recovery Systems** .......... 146
*Mariano J. Savelski, C. Stewart Slater, Eduardo J. Cavanagh*
**(564b) Towards a Complete Solution of Allocation in Life Cycle Inventories** .......... 147
*Rebecca Hanes, Nathan Cruze, Prem K. Goel, Bhavik R. Bakshi*
**(564c) Hybrid Techno-Economic Modeling Tool for Greener Chemicals Supply Chains** .......... 148
*Yuan Yao, Yuan Chang, Fred Thwaites, Eric Masanet*

**(564d) Design Decision-Making Under Uncertainty to Identify Sustainable Integrated Forest Biorefinery Strategies** ........ 149
*Shabnam Sanaei, Paul R. Stuart*
**(564e) MILP-Based Clustering Method for Multi-Objective Optimization: Application to Environmental Problems** ........ 150
*Diego Gabriel Oliva, Gonzalo Guillén-Gosálbez, Laureano Jiménez-Esteller*
**(564f) Emergy Method Analysis On Sludge Residues For Use From Waste-WATER Treatment Plant** ........ 151
*Natalia Andrea Cano, Dario Gallego, Hector Ivan Velásquez*
**(575a) Direct Land Use Change Greenhouse Gas Emissions Associated With Forest-Based Biofuels and Bioenergy Production in Michigan** ........ 154
*Jiqing Fan, Robert E. Froese, David R. Shonnard, Robert M. Handler*
**(575b) Light-Limited Continuous Culture of Micro Algae in a Taylor Vortex Reactor** ........ 155
*R. Dennis Vigil, Bo Kong*
**(575c) Algae Derived Bio-Oil Separation By Supercritical Fluid Fractionation** ........ 156
*Daniel M. Ginosar, Foster Agblevor, Lucia M. Petkovic, Deborah T. Newby, John M. Moses*
**(575d) Enhanced Separation of Algal Biomass By Vibratory Shear Microfiltration for the Sustainable Production of Biofuels** ........ 157
*C. Stewart Slater, Mariano J. Savelski, Pavlo Kostetskyy*
**(575e) Multi-Objective Optimization of Biobutanol Production: Continuous Fermentation Coupled With Vacuum Separation Unit** ........ 158
*Aida Sharif Rohani, Jules Thibault, Poupak Mehrani*
**(575f) Growth, Harvesting, and Modeling of Lipomyces Starkeyi Lipids On Waste Sweet Potatoes** ........ 159
*Melissa Mason, Stephen Dufreche, Mark E. Zappi, Ramalingam Subramanyam, Rakesh Bajpai*
**(575g) A New Approach to Study Selected Aspects of Biomass Oxidation in Supercritical Water** ........ 160
*Javeed Mohammad, Uday G. Hegde, Michael Hicks, Janusz A. Kozinski*
**(600a) Mixed-Conductor Supported Ferrites for Chemical Looping Conversion of Methane** ........ 161
*Yanguang Chen, Fanxing Li*
**(600b) A Novel Mixed Metallic Ni Based Oxygen Carrier for Chemical Looping Combustion** ........ 162
*Hugo de Lasa, Mohammad R. Quddus, Mohammad Mozahar Hossain*
**(600c) Low-Cost Oxygen Carrier Materials for Chemical-Looping Combustion of Biomass** ........ 164
*Georg Schwebel, Gonzalo Vilela Perez, Pavleta Knutsson, Henrik Leion*
**(600d) Fe and Mn-Based Oxygen Carriers for Chemical Looping Combustion** ........ 165
*Saurabh Bhavsar, Amey More, Götz Veser*
**(600e) Experimental and Model Comparison of Mixed Metal Ferrites for H2 Production Using Chemical Looping** ........ 166
*Victoria J. Aston, Brian W. Evanko, Christopher L. Muhich, Jonathan Scheffe, Alan W. Weimer*
**(600f) Sol-Gel Derived Fe2O3-Based, Al2O3 Supported Oxygen Carriers for Hydrogen Production Via a Chemical Looping Cycle** ........ 167
*Nur Sena Yüzbasi, Qasim Imtiaz, Christoph R. Müller*
**(600g) NiO/Ce-?Al2O3 Oxygen Carrier for Chemical-Looping Combustion: Reactivity and Stability** ........ 169
*Shamseldin Mohamed, Shaikh Razzak, Mohammad R. Quddus, Hugo I. de Lasa, Mohammad M. Hossain*
**(616a) Development of Bench-Scaled Adsorption Type Steam Recovery System for Generating High Temperature Steam From Hot Waste Water** ........ 170
*Yoshiaki Kawakami, Yoshio Abe, Kasumi Ito, Kenji Marumo, Toshiyuki Aoyama, Masayuki Tanino, Koichi Nakaso, Tsuguhiko Nakagawa, Yoshinori Itaya, Jun Fukai*
**(616b) Strategic Long-Term Energy Planning for Natural Gas to Liquid Fuels (GTL) Supply Chain Systems** ........ 172
*Josephine A. Elia, Jie Li, Christodoulos A. Floudas*
**(616c) Energy Sustainability: Trade-Offs of Consumption, Efficiency, and Environmental Impact** ........ 174
*Leisha Vance, Tarsha Eason, Heriberto Cabezas*
**(616d) Evaluation of Itev (Air-conditioner Integrated Electric Vehicle) Using in the PV & EV Combined System** ........ 175
*Seishiro Shibata, Kyohei Shimono, Tsuguhiko Nakagawa*
**(616e) Experimental Investigation of Self-Heat Recuperation Using Magnetocaloric Effect for Energy Saving in Thermal Processes** ........ 176
*Yui Kotani, Yasuki Kansha, Masanori Ishizuka, Atsushi Tsutsumi*
**(616f) Energy Diversification in the Middle East: How to Double a Region's Energy Resources** ........ 177
*Massimiliano Kolbe, Zohreh Ravaghi-Ardebili*
**(616g) The Total Site Approach As a Synthesis Tool for the Selection of Valorization Paths in Lignocellulosic Biorefineries** ........ 180
*Michail E. Stefanakis, Konstantinos A. Pyrgakis, Aikaterini D. Mountraki, Antonis C. Kokossis*

**(641a) Algal Biorefinery: Production of Biocrude Oil & Byproducts** ........ 182
*Harvind Kumar Reddy, Thinesh Selvaratnam, Tapaswy Muppaneni, Nagamany Nirmalakhandan, Tanner Schaub, Barry Dungen, Nilusha Sudasinghe Appuhamilage, Peter Lammers, Wayne Voorhies, Shuguang Deng*

**(641b) Biodiesel Production Through Microwave Assisted Transesterification of Microbial Cells** ........ 183
*Yi Cui*

**(641c) Design Optimization By Response Surface Methodology for Continuous Fermentative Production of 1,3 Propanediol From Waste Glycerol By Product of Biodiesel Processes** ........ 184
*Baishali Kanjilal, Iman Noshadi, Nicholas Intoci, Matthew Dowding, William Hale, Brittany Brendel, Ranjan Srivastava, Richard Parnas*

**(641d) Review of Bio-Jet Fuel Conversion Pathways** ........ 185
*Wei-Cheng Wang, Ling Tao*

**(641e) Pellets From Lignocellulosic Materials: Briquetting Process Evaluation** ........ 186
*Yuri Gonzalez, Sahra Rincon, Gabriel Camargo, Fabio Sierra, Carlos A Forero N*

**(641f) Pyrolytic Liquid Yields From Bamboo Species (Bambusa Oldhamii and Bamboo Angustifolia)** ........ 187
*Y. S. Pliego-Bravo, Lidia Cervantes- Pesado, Eusebio BolaÑos-Reynoso, Pedro Gonzalez-Garcia*

**(641g) The Impact of Ethanol and Iso-Butanol Blends on Regulated Emissions, Air Toxics, and Particle Emissions from SIDI Vehicles** ........ 190
*Georgios Karavalakis, Daniel Short, Diep Vu, Thomas Durbin, Akua Asa-Awuku*

**(647a) The Novel Cal-C Process for Single-Stage Removal of CO2, SOx and NOx – Experimental Studies and Process Analysis and Integration** ........ 192
*Niranjani Deshpande, Nihar Phalak, William Wang, L. S. Fan*

**(647b) Carbon Dioxide Absorption Into Aqueous Solution of Hydrophilic Amine-Functionalized Ionic Liquids** ........ 193
*Bihong Lu, Xiangqian Wang, Jiajia Jin, Wei Li*

**(647c) The Effect of Tethering Strategies On the Surface Structure of Amine-Functionalized Ionic Liquids: Inspiration On the CO2 Capture** ........ 195
*Huabin Xing, Yan Yan, Qiwei Yang, Zongbi Bao, Baogen Su, Yiwen Yang, Qilong Ren*

**(647d) A Carbonate Absorption Process With Crystallization-Enabled High Pressure Stripping for Post-Combustion CO2 Capture: A Kinetic Study and Process Development for Bicarbonate Crystallization** ........ 196
*Qing Ye, Yongqi Lu*

**(647e) Carbon Capture and Utilization By Three-Step Aqueous Phase Mineralization** ........ 197
*Azadeh Hemmati, Jie Bu, Jalal Shayegan, Paul Sharratt*

**(647f) Wood Ash Suspensions in Water As a Solvent in CO2 Absorption Processes** ........ 198
*Sandra L. Petho, Wilson Miguel Salvagnini, Maria Elena S. Taqueda*

**(647g) CO2 Chemical Transformation to Valuable Commodities: Toward a Utopia for Fossil-Fuel-Based Economies** ........ 199
*Amin Taheri Najafabadi*

**(664a) Evaluation of Spray Dried Oxygen Carriers Based On Manganese Ore and Ca(OH)2 for Chemical-Looping With Oxygen Uncoupling** ........ 200
*Sebastian Sundqvist, Tobias Mattisson, Henrik Leion, Anders Lyngfelt*

**(664b) Solid Fuel Conversion Via Chemical-Looping With Oxygen Uncoupling (CLOU) Using Co-Precipitation Derived Cu-Mn Oxygen Carriers** ........ 202
*Davood Hosseini, Qasim Imtiaz, Agnieszka Kierzkowska, Christoph R. Müller*

**(664c) Using (MnxFe1-x)2SiO5 As Oxygen Carriers for Chemical-Looping With Oxygen Uncoupling (CLOU)** ........ 203
*Tobias Mattisson, Dazheng Jing, Jasper Van Noyen, Magnus Ryden, Anders Lyngfelt*

**(664d) CO2 Utilization Via Chemical Looping Process** ........ 204
*Vladimir V. Galvita, Hilde Poelman, Guy B. Marin*

**(664e) The Synthesis of Highly Efficient Ca-Based CO2 Sorbents Containing a Hierarchical Pore Structure** ........ 207
*Marcin Broda, Christoph R. Müller*

**(664f) Li4SiO4-Based Novel Solid Fluidizable Sorbent for CO2 Capture:** ........ 209
*Muhammad B.I. Chowdhury, Hugo de Lasa, Mohammad R. Quddus*

**(703a) Isolation, Identification and Characterization of High CO2 Tolerant Microalgae for Efficient CO2 Sequestration From Two Different Locations of India** ........ 210
*Aloke Kumar Ghoshal*

**(703b) Complex Energy System Which Combined Photovoltaic Power Generation With Electric Industrial Vehicle-Case Study of Maniwa SMART Factory-** ........ 212
*Ayako Shimizu, Yuu Notoji, Tsuguhiko Nakagawa*

**(703c) Multi-Objective Optimization of Bioelectricity Supply Chain With Life Cycle Assessment and Social Impact Analysis** ........ 213
*Maxim Slivinsky, Dajun Yue, Fengqi You*
**(703d) Techno-Economic Modeling and Analysis of Energy Storage Options for a Self-Sustained Solar PV Array** ........ 214
*Brian A. Peterson, Joseph D. Smith*
**(703e) Solar Electricity Generation System With Combined Energy Storage and Steam Generation** ........ 215
*Rong Xu, Theodore Wiesner*
**(703f) Thin-Film Composite Membranes for Osmotic Power Generation** ........ 216
*Xue Li, Tai-Shung Chung, Rui Chin Ong*
**(707b) CO2 Capture and Conversion Based On ILs** ........ 217
*Suojiang Zhang*
**(707c) Modeling of CO2 Capture Using Microalgae** ........ 218
*Lidiane Maria Andrade, Maria Anita Mendes, Carlos Alberto Silveira, Cláudio Augusto Oller Nascimento*
**(707d) Biogas Sweetening Via a Carbamate Crystallization Approach** ........ 219
*Shudong Zheng, Qing Liu, Yi He, Yao Shi*
**(707e) Preparation and Absorption Behaviors Study of Doped Li4SiO4 At Low CO2 Concentration** ........ 220
*Zhang Qi, Zhang Sai*
**(707f) Zr Doping Effect On CeO2 As the Catalyst in the CO2 – Ethylbenzene Reaction** ........ 222
*Qian Wang, Wenying Li*
**(707g) Combined Natural Gas Liquefaction and Power Generation Plant With Carbon Dioxide Sequestration** ........ 223
*Sandeep Verma, T. S. Ramakrishnan*
**(721a) Evaluation of Microbial Lipopeptide Biosurfactants for Oil-Spill Remediation** ........ 224
*Mustafa E. Marti, William Colonna, Patra Partha, Haibin Zhang, Chris Green, Gabriel Reznik, Michelle Pynn, Kevin Jarrell, Andy Nyman, Ponisseril Somasundaran, Buddhi Lamsal, Charles E Glatz*
**(721b) Feasibility Study of Production of Methanol and Dimethyl Ether From Flare Gas** ........ 225
*Sho Fukuda, Kazutoshi Chaki, Toshiya Wakatsuki*
**(721c) Abatement of Greenhouse Gases Into Methanol in a Ntp-DBD Reactor - Effects of Experimental Parameters** ........ 227
*Shaik Mahammadunnisa, P Manoj Kumar Reddy, Ch Subrahmanyam*
**(721d) Progress and Economic Considerations for the Biological Production of Triterpene Biofuels From Gases** ........ 229
*Nymul E. Khan, John A. Myers, Ryan Johnson, Alex Rajangam, Eric Nybo, Joe Chappell, Wayne Curtis*
**(727a) Effect of Pore and Particle Size of Oxygen Carriers On Chemical-Looping Combustion and Reforming** ........ 231
*Lu Han, Zhiquan Zhou, George M. Bollas*
**(727b) Thermal Evaluation of Chemical Looping Process for Hydrogen Production Using Bivalent Metal Added Iron Oxide As An Oxygen Carrier** ........ 233
*Wonchul Cho, Kyoungsoo Kang, Chusik Park, Seong Uk Jeong, Doyeon Lee, Myung Won Seo, Chang Hee Kim, Ki-Kwang Bae, Sang Done Kim*
**(727c) Modeling of Cu Oxidation in Air Reactors of Chemical Looping Combustion Systems** ........ 234
*Eli A. Goldstein, Reginald E. Mitchell, Turgut M. Gür*
**(727d) An Overview of Chemical-Looping Reduction in Fixed-Bed and Fluidized Bed Reactors Focused On Oxygen Carrier Utilization and Reactor Efficiency** ........ 235
*Zhiquan Zhou, Lu Han, Shoucheng Du, Ari Fischer, George M. Bollas*
**(727e) Modeling A Scaled 300 kWth Circulating Fluidized Bed Reactor With Barracuda** ........ 237
*Matthew A. Hamilton, JoAnn S. Lighty*
**(727f) The Ohio State University Chemical Looping Processes: 25 Kwth Sub-Pilot Performance Studies for Natural Gas Conversion and 250 Kwth Pilot Unit Developments** ........ 239
*Mandar Kathe, Andrew Tong, Liang Zeng, Dawei Wang, Samuel Bayham, Omar McGiveron, L. S. Fan*
**(727g) Combustion Of Polyurethane By Chemical Looping Process In A Moving Bed Reactor Using Iron-Based Oxygen Carrier** ........ 240
*Young Ku, Ping-Chin Chiu, Hsuan-Chih Wu, Yao-Hsuan Tseng, Yu-Lin Kuo*
**(743a) Towards Engineering Advanced Vanadium Flow Battery** ........ 241
*Liyu Li*
**(743b) Development and Experimental Validation of An Automotive System Model for Slurry-Based Chemical Hydrogen Storage Materials** ........ 242
*Kriston Brooks, Rich Pires, Matt Westman, Jamie Holladay*
**(743c) Deployment of An in-Membrane Micro Fuel Cell for Portable Applications** ........ 243
*Ayokunle Omosebi, Ronald S. Besser*

**(743d) Experimental Study of Solid Fuel Agglomeration and De-Agglomeration Behaviors in the Reactions With Metal Oxide Particles** ..... 244
*Zhenchao Sun, Yuanpeng Ding, L. S. Fan*
**(743e) Development of Fe-Cu Oxygen Carriers for Solid Fuel Gasification and Hydrogen Production** ..... 245
*Nathan Galinsky, Feng He, Fanxing Li*
**(743f) Synthesis and Redox Pathways of MgAl2O4-Stabilized, Fe-Cu-Based Oxygen Carriers for Chemical Looping Water Splitting** ..... 246
*Qasim Imtiaz, Nur Sena Yüzbasi, Marcin Broda, Agnieszka Kierzkowska, Paula Abdala, Wouter van Beek, Christoph R. Müller*
**(743g) Study On Performance of the Novel Steam Generation System Using Water-Zeolite Pair** ..... 248
*Koichi Nakaso, Kazuya Nakashima, Yuuki Tanaka, Yoshiho Iwama, Jun Fukai*
**(774a) A Novel Fluidized Bed Hydrator for Steam Reactivation of Calcium-Based CO2 Sorbent** ..... 250
*Alan Wang, Dawei Wang, Niranjani Deshpande, Nihar Phalak, William Wang, L. S. Fan*
**(774b) Dynamic Modeling of a Solid-Sorbent CO2 Capture System** ..... 251
*Srinivasarao Modekurti, Debangsu Bhattacharyya, Stephen E. Zitney, David C. Miller*
**(774c) Comparative Environmental Life Cycle Assessment of CO2 Capture Systems Based On Ionic Liquid and Monoethanolamine Solvents** ..... 252
*Reza Farahipour, Arunprakash T. Karunanithi*
**(774d) Whole-Chain CCS System Modelling: Enabling Technology to Help Accelerate Commercialisation and Manage Technology Risk** ..... 253
*Mark Matzopoulos, Alfredo Ramos*
**(774e) Optimization of CO2 Capture, Utilization and Sequestration (CCUS) Supply Chain Networks** ..... 254
*M. M. Faruque Hasan, Fani Boukouvala, Christodoulos A. Floudas*
**(808a) CO2 Capture By Gas Hydrate Formation: Combined Effect of Thermodynamic and Kinetic Promoters** ..... 256
*Nagu Daraboina, Nicolas von Solms*
**(808b) CO2/Brine/Rock Interactions Under CO2 Sequestration Conditions** ..... 257
*Yee Soong, Bret H. Howard, S. W. Hedges, Dustin McIntyre, Robert P. Warzinski, Gino Irdi, Igor Haljasmaa, Dustin Crandall*
**(808c) Novel CO2 Capture and Carbonation Process With Waste Concrete** ..... 260
*Daiki Shuto, Akihiro Yamasaki, Teruhisa Hongo, Atsushi Iizuka*
**(808d) Direct Capture and Conversion of CO2 From Flue Gas Via Ex-Situ Mineral Carbonation Using Chemical and Biological Catalysts** ..... 261
*Edward J. Swanson, Patrick V. Brady, Ah-Hyung Alissa Park*
**(808e) A Green Approach to Mitigate CO2 Using Mixed Microbial Culture** ..... 262
*Smita Raghuvanshi, Suresh Gupta, Somesh Mishra*
**Author Index**

**(32a) Overview Of Materials In The Built Environment**

Judkoff, R.,

Originally presented on: 11/4/2013 8:30:00 - 8:52:00

Materials advances could help to reduce the energy and environmental impacts of buildings. Globally, buildings consume 30%-40% of primary energy and account for 25%-33% of $CO_2$ emissions. This talk explores some foundational concepts about how energy is used in buildings, and also highlights areas where materials advances would be most beneficial from an energy perspective.

**(32b) Energyplus Model-Based Predictive Control (EPMPC) Using Matlab/Simulink and Mle+**

Zhao, J., Center for Building Performance and Diagnostics
Lam, K. P., Center for Building Performance and Diagnostics
Ydstie, B. E., Carnegie Mellon University
Originally presented on: 11/4/2013 8:52:00 - 9:14:00

This paper proposes the EnergyPlus Model-based Predictive Control (EPMPC) system to control under floor air distribution (UFAD) in a multi-zone two stories open office building – The Center for Sustainable Landscapes – at Phipps Conservatory in the city of Pittsburgh. The building is equipped with a central air-handling unit with ground source heat pump as the cooling and heating source. The UFAD is used for the open office, conference rooms and other occupied spaces. Ceiling-based air distribution system is used for service spaces, such as restrooms, mechanical rooms, and storage rooms. The ASHRAE 90.1-2007 baseline and design phase whole building models are created using DesignBuilder and EnergyPlus program under the "Design-Build-Operate Energy Information Modeling (DBO-EIM)" infrastructure. The key of the DBO-EIM infrastructure is to use the detailed design phase energy model throughout the entire building life cycle by validating and adapting model assumptions during construction, commissioning and operation periods. Therefore, the model is expected to be "accurate" and easy-to-use for model-based building control systems in the daily operation to optimize energy and comfort. In this paper, the design phase model is applied in both the baseline and the EPMPC control. The EPMC controller includes both continuous and discrete variables and it is not possible to use optimization algorithms that rely on gradient search. In order to solve the optimization problem we developed an exhaustive search optimization algorithm within the Matlab/Simulink environment via maximum likelihood estimation+, co-simulation tool. In two weeks of winter and summer simulation using *typical meteorological year 3* weather file, comparing to the baseline logic control, which is also implemented using the same co-simulation method, the EPMPC can save HVAC operation energy by 18.9%. The occupant *predicted mean vote* value is as good as the baseline if not better. The benefits of using the EnergyPlus model over the simplified first principle or identified linear model include: (1) it can model complex buildings and systems; (2) it can be continuously updated with little cost and effort; (3) it requires less mathematics skills to manage in daily operation once the system has been developed.

**(32c) The Sol-Char Toilet: Concentrated Solar Power Transported By Fiber Optics for Pyrolysis of Human Waste to Biochar**

Fisher, R. P., University of Colorado at Boulder
Lewandowski, A., National Renewable Energy Lab
Yacob, T. W., University of Colorado at Boulder
Mahoney, R. B., University of Colorado at Boulder
Mejic, D., University of Colorado at Boulder
Oversby, C., University of Colorado at Boulder
Beck, S., University of Colorado at Boulder
Ward, B. J., University of Colorado at Boulder
Kearns, J., University of Colorado at Boulder
Summers, R. S., University of Colorado
Linden, K. G., University of Colorado at Boulder
Weimer, A. W., University of Colorado at Boulder
Originally presented on: 11/4/2013 9:14:00 - 9:36:00

Currently, 2.6 billion people on the planet do not have access to adequate sanitation and many communities in the developing world have neither the facilities nor the resources to properly treat human waste, leading to open defecation posing enormous public health risks as well as issues related to individual dignity and safety. Sanitation is not a problem unique to developing countries – modern sanitation processes are also unsustainable due to their often massive requirements for energy and water. The Sol-Char Toilet, a solar waste treatment technology developed at the University of Colorado Boulder, takes advantage of our most abundant renewable resource: the sun. In the Sol-Char Toilet, concentrated solar power (CSP) is transmitted through fiber optic cables to drive pyrolysis of human waste, converting a hazardous and unusable material to biochar. Biochar is a carbonized solid product of thermal biomass decomposition consisting largely of condensed aromatic zones that when applied as a soil amendment imparts agronomic benefits and is recalcitrant over a long timescale ($10^2$ – $10^3$years). Biochar presents a solution relevant to the world today: reducing greenhouse gas emissions by sequestering carbon in biochar for several hundreds of years or even millennia. The over-reaching goal of the project is to develop a cost-effective, self-contained, and sustainable sanitation process by which human waste can be converted to useful products without connection to water, sewer or electricity.

Sunlight is concentrated to nearly 2000 Suns by a parabolic mirror, directed into a fiber optic bundle, and routed to a solar hood that directly irradiates a pyrolysis reactor. Two identical reactor vessels are switched between a pyrolysis location under the solar hood and a toilet location via a rotating carousel mechanism, thus achieving a semi-continuous pyrolysis process during daylight hours. The Sol-Char Toilet prototype represents a novel application of CSP to the areas of environmental, sanitation, and reaction engineering. This presentation will focus on the design and operation of the 1 kW Sol-Char Toilet prototype designed and constructed at the University of Colorado at Boulder.

**(32d) Building-Integrated Active Modular Phytoremediation System**

Dyson, A., Rensselaer Polytechnic Institute/Center for Architecture Science and Ecology
Aydogan Akseli, A., Rensselaer Polytechnic Institute
Vollen, J., Rensselaer Polytechnic Institute
Nyman, M., Rensselaer Polytechnic Institute
Bird, J.,
Originally presented on: 11/4/2013 9:36:00 - 9:58:00

Contemporary construction materials and mechanical building ventilation systems types are complicit in the degradation of indoor air quality (IAQ) and have become principal contributors to public health concerns in developed countries. These challenges are further exacerbated by the increasing cost of energy that is required to maintain healthy levels of IAQ in these buildings. Addressing these issues will require not only innovative technologies and approaches, but also deep collaborations across several disciplines.

This research stems from an interdisciplinary framework for the development of a building-integrated phytoremediation system using bio-mechanical techniques to 'scrub' toxins from the air and maintain desired psychometric profiles for IAQ by integrating modules containing hydroponic-supported plant materials into heating, ventilating and air conditioning (HVAC) systems. This research is multi-dimensional and interdisciplinary in scope, with the precise goal of integrating lab-based testing protocols to reiterative architectural systems design parameters.

There are three scopes that will be presented within this research: 1.Parametric framework for information exchange across disciplines; 2.Acqusition of scientific data on the sorption of VOCs (*e.g.,* formaldehyde and benzene) to growing media and granular activated carbon, removal efficiency of the selected VOCs from air utilizing the building-integrated phytoremediation system and the effect of phytoremediation as a bio-mechanical technique to remove formaldehyde and benzene from indoor air and; 3. Co-design of an integrated bio-mechanical HVAC system at building-scale. A series of experiments were undertaken to determine the performance parameters for architectural integration, including: the toxin removal capacity of various growing media and plant species; the impact of module morphology on contaminant concentration levels and air flow rates; and the rates of remediation by constituent components within plant leaves and the root rhizosphere. The conclusions from these experiments show that the building-integrated active modular phytoremediation system (AMPS) could have a significant impact in contributing to remediation strategies addressing poor IAQ in buildings and energy reduction through improved HVAC performance.

## (32e) Integrating Technological and Ecological Design for Sustainable Buildings On a University Campus

Gibbemeyer, E. L., The Ohio State University
Bakshi, B. R., The Ohio State University
Originally presented on: 11/4/2013 9:58:00 - 10:20:00

### Importance

The Ohio State University has pledged to become carbon neutral by 2050, an ambitious and currently out of reach goal. Technological systems often find the task of carbon neutrality to be challenging and frequently impossible. Therefore, we look to nature and ecosystems as examples of systems that have maintained sustainable levels of carbon, as well any other resource we care to examine. Thus, instead of merely optimizing a technological system to have minimum carbon emissions, we also consider the surrounding ecosystem and optimize the entire system to minimize carbon in what we call "techno-ecological networks." For example, in optimizing a building, our technological system, we can optimize it to be as close to carbon neutrality as possible; however, by simply planting additional trees outside of it, we can further reduce its carbon footprint by accounting for the carbon these trees will sequester. Including the ecosystem in our network allows us to reach a smaller carbon footprint that was previously infeasible when only looking at the technological components of our system.

### Originality

While there have been many studies evaluating and optimizing the Ecological Footprints of buildings, ecosystems are usually not included in the evaluation. This means that any optimized solutions found in these studies could be further improved by also including the surrounding ecosystems. We also plan to include behavioral variables such as choosing a thermostat set point. Previously, we have incorporated these ideas into a simplified residential system; in this project, our techno-ecological network will be the real system of a segment of the OSU Columbus Campus that will include buildings and their surrounding green space. This real system will allow us to continue testing the effects of including ecological and behavioral variables into this optimization framework of minimizing carbon footprint, water footprint, and cost.

### Methods Employed

We will use software developed by the DOE to model our campus buildings, and the models will be verified using data collected by the OSU Energy Services and Sustainability office. We then will begin making changes to technological variables to see what retrofitting options would lower our carbon and water footprints, as well as adding in the ecological variables through biophysical models including InVEST and CENTURY. Once we have developed all the different variables, we will use optimization software to solve for scenarios that minimize carbon, water, and cost.

### Results

Because our optimization is multi-objective, we will not simply obtain a single solution; we will find a Pareto optimal set of solutions. However, we anticipate some variables to lead to "win-win" scenarios, such as installing a rainwater capture system to lower water footprint and also lower water utility costs. In finding "win-win" scenarios, we hope to find opportunities for changes that can be made and allow this study to turn into a true "living laboratory" where students can become more ecologically conscious. Nonetheless, for other variables, we will require a judgment call on the relative importance of our objectives if we want to be able to select a "best" scenario that we could implement on campus.

**(32f) Energy Conservation Approach to Cooling Process By Employing Preheat Tank and Chilled Water Tanks**

Peters, D. R. W., University of Alabama at Birmingham
Li, Z., University of Alabama at Birmingham
Originally presented on: 11/4/2013 10:20:00 - 10:42:00

Energy conservation has been a hot topic worldwide. With the rapid growth of population and urbanization, the energy demand is increasing annually. According to the 2005 *Annual Energy Review* compiled by the Energy Information Administration (EIA) within the U.S. Department of Energy (DOE), the energy consumption in the United States including transportation, residential, and commercial end uses, shows a trend of increasing during the period of 1970 to 2005. Over the years from 1970 to 2004, total commercial end uses have an annual growth rate of 2.6 percent, and electricity consumption accounts for 76.2% of total commercial energy consumption. The HVAC system is an important process to provide comfortable temperatures, but demanding a large amount of energy and mainly contributing to the peak electricity consumption. The chilled water tank is a technology that can dramatically reduce the peak energy consumption with installation in traditional air conditioning systems, especially for the hot and humid climates. The preheat tank is a device used for heat recovery purposes that absorbs the heat from air flows to decrease the air temperature entering the chilled water tank and enhance the performance of cooling with even less energy requirements.

Our research utilizes the preheat tank and chilled water tank in commercial and conventional air conditioning systems to achieve energy conservation opportunities in Birmingham, Alabama. The preheat tank was installed to exchange heat with hot air, and the heated water can be used for lavatory and other purposes. The chilled water tank was set up to provide cooling with less energy consumption. The Generalized Fluid System Simulation Program (GFSSP) software was utilized to analyze the air conditioning system involving the heat recovery tank and chilled water tank. In addition, experiments were conducted in a small building at the University of Alabama at Birmingham campus with an installed small preheat tank and chilled water tank, respectively, to provide data needed for software programming. The energy consumption of the modified HVAC system was monitored to check the energy conservation rate. Some factors that may affect the performance of the modified system, include ambient temperature, humidity, water temperature in the tank, air conditioning set temperature, and dew point. The variables were recorded and analyzed in regression models to determine which factor(s) are significant for energy conservation. Moreover, covariance analysis was performed to determine the relationship between these factors. The results are expected to provide energy conservation for air conditioning systems, especially during the peak hours of energy consumption in hot and humid summer climates which is the typical weather condition in Birmingham, Alabama. Finally, the payback period was estimated by comparing energy bill savings and capital investment. The research can be extended to other southern states that experience hot and humid climates, such as Texas, Arizona, Georgia, and Florida.

**(32g) Operation of Residential Hybrid Renewable Energy Systems: Integrating Forecasting, Optimization and Demand Response**

Wang, X.,
El-Farra, N. H., University of California, Davis
Palazoglu, A., UC Davis
Originally presented on: 11/4/2013 10:42:00 - 11:04:00

Hybrid renewable energy systems, which consist of solar, wind and other energy generation and storage units, have been widely researched in recent years owing to the reduction in traditional fossil resources and increase in the world-wide energy demand. This paper focuses on the optimal design and operation of a standalone community-scale hybrid renewable energy system comprising photovoltaic (PV) cells, wind turbines, diesel generators and storage units such as batteries. A detailed modeling of each system component is presented. The demand-side management as well as the performance of the generation system are considered to guarantee that the electricity demand of the consumer is satisfied and the overall system economic and environmental cost is minimized.

To optimally allocate electricity generation among different units, a precise prediction of the future electricity consumption and the available solar and wind resources is necessary. Both the day-ahead and real-time predictions are made to provide sufficient information for further optimization. For the day-ahead prediction, historical demand as well as solar radiation and wind speed data are used as the basis for the classification of day types, which are determined by the temperature, humidity, seasonality and other related characteristics. During the day-ahead operational management, a general electricity usage and a generation profile are developed to realize the preliminary dispatch strategy. For the real-time prediction, autoregressive and polynomial fitting methods are adopted to facilitate time-series analysis and forecasting. The online forecasting model is able to consider the updated data and modify the overall prediction and optimization.

On the demand side, a demand-responsive scheme is integrated with the operation of power generation. Among all the residential electricity usage, some are rigid in demand which cannot be changed or shifted such as lighting and cooking while others are flexible loads including washing machines, clothes dryers and vacuum cleaners, which are the key factors to be considered to manage the demand side. These loads can be arranged to reshape demand profiles to match renewable output better.

In the optimization stage, the method of model predictive control (MPC) is adopted by considering real-time optimization to minimize multiple objectives over a receding horizon, the most significant of which is to minimize the generation cost within the time window considered. The environmental influences are also accounted for by the indicators of equivalent cost. As constraints, each energy resource has its maximum generation capacity, as well as minimum output level, which decides the operation range of the generation unit. Limits of ramping rate of each resource are also posed to avoid damage caused by excessively rapid change of power output. The formulation is simplified to a convex optimization problem. At each time interval, only the first set of power references calculated by the optimization are implemented. The optimal allocation between different resources drives the generation to meet electricity requirement of the demand side, with minimum operating cost and environmental impact. Different strategies to realize the receding-horizon method are applied and the results are compared. It turns out that the length of the moving horizon has a critical impact on the optimal allocation of electricity generation among the available resources. Proper selection of the horizon allows for an efficient and optimal operation of the energy system.

The strength of the proposed strategy is that it is able to achieve global optimization over a certain time period and at the same time the users are able to obtain sufficient information of the future conditions and thus make rational decisions about electricity management. Compared with traditional operation of energy systems, the on-line process that combines real-time prediction, optimization and demand-responsive scheme will improve system efficiency and take better advantage of all available resources.

We will show several simulation and optimization case studies that demonstrate the effectiveness and applicability of the proposed receding horizon scheme. Future work to connect the current system to the electrical smart-grid to realize large-scale application will also be discussed.

**(36a) Green Chemistry, Green Engineering, Green Economy At US EPA 's Sustainable Technologies Division**

Leazer, J., U.S. Environmental Protection Agency

Originally presented on: 11/4/2013 8:35:00 - 8:50:00

Green Chemistry and Green Engineering Research in the sustainable Technologies Division of USEPA's National Risk Management Research Laboratory is focused on reducing the adverse human health and environmental impacts of the synthesis of economically important chemicals and chemical processes. Our research focuses on several categories including:

1) Sustainable Materials Management, which accounts for ecological scarcity and resource efficiency;

2) Green Chemistry, which focuses on the minimization/elimination of human health and environmental impacts as well as resource efficiency;

3) Green Engineering, which focuses on energy and resource conservation;

4) Sustainability Measurement, which allows us to determine whether a system is becoming more sustainable or not;

5) Life Cycle Assessment, ensuring minimal impact across the entire life cycle of a product/good/service and

6) Supply Chain Design, which looks at the optimization and increased sustainability of the entire supply chain of a product.

These research areas have the common goal of green production of economically viable chemicals and chemical processes.

**(36b) Sustainability Practice in Industry**

Powell, J. B., Shell Westhollow Technology Center

Originally presented on: 11/4/2013 8:50:00 - 9:05:00

This seminar will address how sustainability is implemented in industry, with real world examples from the energy and chemical processing industries. Topics will include making the case for sustainability, sustainability metrics and reporting, the importance of considering sustainability early during new process or product development, and how industry may be organized to incorporate sustainability principles into decision making. The importance of external stakeholder engagement in project planning and execution will be addressed.

In industry, sustainability is viewed as a key enabler for license to operate, and can be seen as a driver for much of the new innovation and development in our profession. Competency in life cycle inventory and overall environmental footprint and social impact assessment has become increasingly important in assessing preferred opportunities for technology development and business deployment. Demand for excellence in safe and responsible implementation and performance will continue strong for the developed world, and dissemination of best technologies and practices to developing economies will be critical for addressing global sustainability needs and environmental challenges.

**(36c) Green Engineering and Sustainability in the Chemical Engineering Curriculum**

Shonnard, D. R., Michigan Technological University
Allen, D. T., The University of Texas at Austin
Nguyen, N., U.S. Environmental Protection Agency
Austin, S., Chemical Engineering Branch, US EPA
Originally presented on: 11/4/2013 9:05:00 - 9:20:00

A green economy will be well-served by an educated chemical engineering workforce firmly grounded in green and sustainable engineering concepts. The goal of this presentation is to outline updated green engineering and sustainable engineering content for undergraduate and graduate chemical engineering education. The work of chemical engineers is central to the development of sustainable chemical and energy production systems and to meet global human needs. Incorporating environmental issues and sustainability into the design of engineered products, processes, and systems generally means evaluating environmental, economic, and social performance. This presentation will summarize two approaches to incorporate green and sustainable engineering into undergraduate chemical engineering education; **i.** a stand-alone elective course, and **ii.** modules for required core chemical engineering courses. A detailed overview of chapter-by-chapter content will also be presented with tie-ins to core courses.

**(36d) Driving Private Sector Adoption of Greener Processes and Technologies**
Lindsay, A., Author: Abby Lindsay, American University, Doctoral Candidate
Originally presented on: 11/4/2013 9:40:00 - 9:55:00

In addition to advanced environmental technologies and improved chemical processes, a green economy will require widespread adoption by the private sector. This presentation will outline ways in which greener practices are spread to a wider number of companies through focusing on experiences throughout Latin America. Most larger companies with extensive resources and access to financing have realized the environmental and cost benefits to reducing energy and water use, decreasing pollution, and increasing water and materials efficiency. Micro, small, and medium-sized enterprises (MSMEs) can benefit just as much, but often lack the awareness, financial resources, or skills to increase their efficiency. Given that the majority of companies are MSMEs, reducing the resource-use and pollution intensity of these companies is critical to a green economy. This presentation will look at factors that drive the diffusion of environmental processes and technologies, and determinants required for successful adoption, such as adequate engineering education systems, availability of financing, and policy incentives.

**(36e) Minimization of the GHG Emissions At the Macroeconomic Level Via a Multi-Objective Input-Output Approach: A Case Study of the EU-25 Economy**

Cortes, D., University Rovira i Virgili
Ruiz-Hernandez, A., University Rovira i Virgili
Guillen-Gosalbez, G., Universitat PolitÃ©cnica de Catalunya
Llop, M., Universitat Rovira i Virgili
Guimera, R., Universitat Rovira i Virgili
Sales-Pardo, M., Universitat Rovira i Virgili
Originally presented on: 11/4/2013 9:55:00 - 10:10:00

**Minimization of the GHG emissions at the macroeconomic level via a multi-objective input-output approach: A case study of the EU-25 economy**

ABSTRACT

D. Cortes-Borda[a], A. Ruiz-Hernández[a], G. Guillen-Gosálbez[a], M. Llop[b], R. Guimerà, M. Sales-Pardo.

a.Departament d'Enginyeria Química, Universitat Rovira i Virgili, Av. Països Catalans, 26, Tarragona E-43007, Spain.

b.Centre de Recerca en Economia Industrial i Economia Pública (CREIP), Universitat Rovira i Virgili, Av. Universitat 1, 43204 Reus, Spain.

Global warming has grown significantly in the recent past due to the increasing atmospheric concentration of anthropogenic greenhouse gasses (GHGs). As a result, worldwide environmental policies are placing global warming mitigation as a top priority in the nations' environmental agendas. That is, most of the National governments are adopting policies that aim to minimize GHG emissions.

Unfortunately, in today's globalized markets designing effective environmental policies is challenging because international trade establishes channels through which impacts are imported and exported among nations, overshadowing the main source of pollution into an intricate trading network. Therefore, the design of these policies requires robust environmental assessment methods capable of capturing the complexity of the existing international supply chains of goods/services .

Methods based on extended environmental input-output (EEIO) models are particularly suitable to address global warming at a global scale, since they can assess macroeconomic connections between productive sectors of different nations and the associated environmental impact. This information can be used to allocate the corresponding environmental responsibility. EEIO models are flexible, transparent and accurate enough to perform life cycle assessment (LCA) studies. Hence, LCA-based EEIO models translate the economic output of a given economy into a tangible environmental impact.

By using macroeconomic EEIO models, we can identify a wide range of alternatives for environmental savings, as it is possible to reduce each sector's output (and therefore its associated GHGs emissions) to a different extent through taxation and/or economic penalties. In this context, systematic optimization tools are especially helpful to identify the most promising alternatives considering economic and environmental concerns.

We have developed in this work a systematic approach of this type for reducing the environmental impact of an economy that is based on the combined use of EEIO models and multi-objective linear programming techniques. This tool has been applied to the European Union (EU-25) economy in order to find solutions that simultaneously maximize the total output and minimize the GHGs emissions. Hence, in our approach the problem of identifying key economic sectors that should be given higher priority is posed in mathematical terms as a bi-criteria LP problem that seeks to optimize simultaneously the total European gross domestic product and the associated life cycle GHGs simultaneously. Our study makes use of the E3IOT database, which covers 487 EU economic sectors.

Numerical results show that substantial reductions in global warming potential can be achieved by imposing taxes on sectors such as energy, mineral extraction, and meat and its derived products. By applying this wide-scope assessment, we find potential alternatives for reducing GHG emissions (that are not obvious at first sight) that can assist policy-making.

**(36f) Economic Analysis and Optimization of Carbon Capture Storage Pipeline Transport Network System**

Dan, S., Seoul National University
Kang, B. J.,
Cho, S.,
Yoon, E. S., Seoul National University
Originally presented on: 11/4/2013 10:10:00 - 10:25:00

Carbon Capture & Sequestration (CCS) is the effective means that can be used to reduce CO2 emissions. Among the CCS network system, a pipeline transport network is the least cost element of the full CCS network system, but the most planning part during the development of the CCS network system.

The research presented here based on the physical properties of CO2 and Co2 containing mixtures calculated. The methodology of this research consists of four steps: Optimal CO2 pipeline design, CCS transport network system modeling, Cost evaluation for model, Selection of the optimal CCS transport network system.

Objective of this research is to find the optimal proposal of CCS pipeline transport network system. This research proposes the optimal engineering CO2pipeline design and CCS pipeline network design for transporting CO2 using HYSYS 7.3 simulator.

**(36g) Economic Development Through Entrepreneurship and Technology Innovation: Bringing Together Industry, Academia, and Government**

Anid, N. M., New York Institute of Technology

Originally presented on: 11/4/2013 10:25:00 - 10:40:00

The School of Engineering and Computing Sciences (SoECS) of New York Institute of Technology (NYIT) is launching a new Entrepreneurship and Technology Innovation Center (ETIC). The ETIC will consolidate and expand the School's ongoing industry-academic partnerships, foster innovation and promote collaborations between industry, the academic community, professional organizations, and government. The Center will support the NY region's economic development by focusing on three critical areas Health Analytics & Bioengineering, IT & Cybersecurity, and Energy and Green technologies. This paper will discuss a case study of a green economic development initiative that involves chemical engineers: the creation of a microgrid pilot installation that utilizes various storage technologies. Process economics and financing via a public private partnership and seed government funding will be presented. The goal is to infuse the engineering curriculum with real-world problems, and to show how project-based learning, teamwork, and a collaborative environment and direct contact with practitioners, instill in students the key competency and entrepreneurship skills required in today's rapidly evolving, globalized economy.

**(37a) The Carbon Capture Simulation Initiative (CCSI): Computational Tools To Accelerate Commercial Development**

Miller, D. C., U.S. Department Of Energy, National Energy Technology Laboratory
Originally presented on: 11/4/2013 8:30:00 - 8:55:00

The U.S. Department of Energy's Carbon Capture Simulation Initiative (CCSI) is developing, demonstrating and deploying advanced computational tools and models to accelerate the development of next generation technologies, specifically the development of cost effective carbon capture technologies. This CCSI Toolset will (1) enable promising concepts to be evaluated more quickly based on an optimized process, (2) reduce the time for design and troubleshooting by integrating process and device scale simulations to better predict performance and more effectively resolve issues, (3) quantify technical risk to enable more focused scale up activities, and finally, (4) stabilize the costs during commercial deployment by having a more comprehensive understanding of the process and underlying behavior of the system. This paper will provide an overview of the program and highlight recent accomplishments applying the tools to potential carbon capture technologies.

**(37b) The Four Dimensional Product: Integration Over Time Is the Only Way to Understand Sustainability**

Banholzer, W., Dow Chemical Company
Jones, M. E., The Dow Chemical Company
Originally presented on: 11/4/2013 8:55:00 - 9:20:00

Sustainability is a question of time, integrating knowledge of production, use and end-of-life. Far too often, hype has surpassed reality, boundary conditions have been neglected and actions taken that don't truly lead to a sustainable future. Singular focus on attributes like renewable content or biodegradability gives the wrong answer. Quality engineering forms the foundation for any meaningful discussion of sustainability, providing insights into energy and material flows, energy return, and thermodynamic constraints. The chemical industry is increasing resource and energy efficiency through innovation in both products and processes. Our mastery of materials science enables unique energy producing and energy saving products, products that improve sustainability by being significantly better than the next best alternatives during use.

**(37c) Maximization of Energy Efficiency of a Combined Heat and Power Plant**

Edgar, T. F., The University of Texas at Austin
Powell, K., The University of Texas at Austin
Kim, J., The University of Texas at Austin
Kapoor, K., The University of Texas at Austin
Originally presented on: 11/4/2013 9:20:00 - 9:45:00

Energy efficiency is receiving increased attention as a way to reduce the use of fossil fuels and the resulting production of greenhouse gases. Automation, process control, and real-time optimization are key technologies to operate plants in the most efficient way. A combined heat and power plant at the University of Texas at Austin has been studied in order to reduce its carbon footprint as well as reduce costs. The steam power plant has been modeled using steady state mass and energy balances. Chillers have also been modeled so that their efficiencies can be maximized. There is also a thermal storage tank for chilled water to be used during periods of high demand. Forecasting of loads as well as the effect of weather have been incorporated into the real-time optimization approach. The effects of time of day pricing of power and use of demand response techniques to flatten demand curves are discussed. Usage of thermal energy storage along with automatic control gives additional degrees of freedom to deal with dynamic power conditions.

**(37d) Pan American Issues on Biofuel Sustainability**
Shonnard, D. R., Michigan Technological University
Originally presented on: 11/4/2013 9:45:00 - 10:10:00

Biofuels are in rapid development in many Pan American countries, yet the nature and extent of impacts on social and environmental systems from large-scale biofuel development are not well understood. This talk will explore sustainability issues across a broad set of impacts and in different Pan American countries and will begin to outline important research gaps. A future research agenda to overcome these knowledge gaps would help establish a new and growing biofuels industry onto a sustainable path.

**(50a) Crosslinked Anion Exchange Membranes for Redox Flow Batteries**

Jung, M. S., Illinois Institute of Technology
Parrondo, J., Illinois Institute of Technology
Arges, C., Illinois Institute of Technology
Ramani, V., Illinois Institute of Technology
Originally presented on: 11/4/2013 8:30:00 - 8:55:00

All-vanadium redox flow batteries (VRFB) have been very popular as they use the same electrochemical species in both electrode chambers and therefore they do not suffer from electrolyte cross contamination. A core component of any RFB is the electrolyte membrane (PEM or AEM) used to separate the positive and negative electrolytes. The key requirements for the separator are good chemical and electrochemical stability, good mechanical stability, high ionic conductivity, and low permeability for the vanadium ions. Nafion® membranes have been the most popular separators in RFBs because of their excellent oxidative stability and high proton conductivity, however, they exhibit high vanadium crossover. Anion exchange membranes (AEMs) are ideal candidates as RFB separators as they have much lower vanadium ion crossover than proton exchange membranes (10-100 times less) due to Donnan exclusion. Moreover, the reduced penetration of $VO_2^+$ into the AEM minimizes chemical degradation.

We selected poly (2,6-dimethyl 1,4-phenylene) oxide (PPO) based AEMs as battery separators as they possess good conductivity, low water uptake and good mechanical stability. PPO is commercially available, relatively inexpensive, and can be readily modified by attaching bromine to the backbone to induce a phase-separated morphology. PPO can be easily functionalized into an AEM by bromination followed by reaction with a tertiary amine. In this talk, we will discuss the effect of polymer crosslinking on the conductivity, water uptake, vanadium (IV) permeability, ultimate tensile stress, glass transition temperature and storage modulus of PPO- based AEMs. Polymer crosslinking has been performed by the reaction of brominated PPO (Br-PPO) with tetramethylhexanediamine (TMHDA) in liquid phase. Non crosslinked AEMs have been prepared by reaction of the Br-PPO with quinuclidine (PPO-ABCO$^+$) and trimethylamine (PPO-TMA$^+$). Nuclear Magnetic Resonance techniques [$^1$H-NMR, $^{13}$C-NMR, correlation spectroscopy (COSY), and $^1$H-$^{13}$C heteronuclear multiple quantum correlation (HMQC) spectroscopy] have been used to investigate the polymer structure before and after exposure to $VO_2^+$ solution (over time) and to identify any possible degradation products and mechanisms. Crosslinked membranes will be tested (and compared against Nafion® and non crosslinked AEMs) in a vanadium redox flow battery to evaluate its performance and durability. Postmortem analysis of the membranes via NMR, conductivity/IEC measurements, and mechanical testing will be performed to unearth key degradation modes.

**(50b) Challenges to Implementation of High-Rate Energy Storage in Military Applications**

Heinzel, J., Naval Sea Systems Command - Philadelphia
Ostanek, J.,
Hoffman, D., Office of Naval Research
Peek, I., Naval Sea Systems Command - Philadelphia
Originally presented on: 11/4/2013 8:55:00 - 9:20:00

Future military platforms will require more extensive use of electronic power systems to achieve required performance levels. Energy storage is becoming a key enabler for power systems operating with high levels of transient power requirements and supporting fuel-efficiency operations while maintaining capability. High density energy storage which operates continuously at high charge and discharge rates to support these loads under the combination of finite space, high power and energy requirements, stringent safety characteristics and wide upper temperature ranges drive the requirement for advancement in a multiphysical mix of electrical, chemical, thermal, safety and controls elements, which must be considered and have development performed to enable widespread implementation. These will be discussed from the perspective of requirements and considerations for system design and engineering for applications associated with high rate support of transient loads, power quality and optimization of generator and electrical architecture dynamics.

**(50c) Novel Pt/TixM1-xO2 Nanocatalysts for Fuel Cells**

Hwang, B. J., National Taiwan University of Science and Technology
Ho, V. T. T.,
Pan, C. J., National Taiwan University of Science and Technology
Su, W. N.,
Originally presented on: 11/4/2013 9:20:00 - 9:45:00

This work presents a new approach by exploring robust non-carbon $Ti_xM_{1-x}O_2$ as a novel functionalized co-catalytic support for Pt. This new approach is based on the novel nanostructure of the $Ti_xM_{1-x}O_2$ supports which holds strong metal-support interactions (SMSI) between $Ti_xM_{1-x}O_2$ and Pt that can not only modify surface electronic structure of Pt, evidenced from a shift in the d-band centre of the surface Pt atoms but also provide dual reaction sites for methanol oxidation. Furthermore, another benefit of $Ti_xM_{1-x}O_2$ is the extremely high stability of Pt/ v during potential cycling, which is attributable to the SMSI between Pt and $Ti_xM_{1-x}O_2$. Interestingly, $Ti_xM_{1-x}O_2$ can be fabricated as a much thinner catalyst layer, resulting in significantly improved the mass transport kinetics and performance of the resulting membrane-electrode-assembly (MEA).The new approach presented in this work opens a reliable path to the discovery of advanced concepts that may lead to the design of new catalyst materials that can replace the traditional catalytic structures and motivate further research in this field.

## (50d) A Tunable Hierarchical Porous Carbon Structure for Studying Electrochemical Energy Conversion

Chan, K. Y., The University of Hong Kong
Yang, C.,
Originally presented on: 11/4/2013 9:45:00 - 10:10:00

### A Tunable Hierarchical Porous Carbon Structure for Studying Electrochemical Energy Conversion

Chunzhen YANG and Kwong-Yu CHAN*, Department of Chemistry, The University of Hong Kong, Pokfulam Road, Hong Kong

Abstract

A well-defined porous carbon structure with tunable parameters provides a platform for systematic study of transport and confinement effects on electrochemical conversion. Synthesized via silica template techniques, the hierarchical structure has a hollow core within a mesoporous shell (HCMS) with uniform pore size, shell thickness, hollow core diameter, as shown in Fig. 1.
By varying the synthesis parameters, the structural parameters can be tuned to study

1) combined effects of surface area, mesopore size, macropore size, and hierarchy of porosity on ionic transport and adsorption in electrochemical capacitance behavior;

2) how distribution of platinum or other electrocatalyst nanoparticles in the porous structure affect catalytic faradaic reactions;

3) how distribution of mixed metal nanoparticles such as PtRu and varying of Pt and Ru content in the porous structure affects faradaic reactions such as methanol oxidation; and

4) the effectiveness of a mathematic model in analyzing electrochemical conversion in porous structures.

Electrochemical capacitance is studied with a series with meso-shell thickness stepwise increased from 0, 25, 50 to 100 nm while keeping an identical 330 nm hollow core and mesopore of 3.9 nm.[1] A thicker shell has a higher surface area with a proportional increase of electrochemical capacitance which however, can only be fully realized at low scan rates/currents. At high currents, ionic transport limits the electrochemical capacitance of a thick mesoporous shell. Electrochemical impedance spectra (EIS) obtained on the family of HCMS carbon structures were fitted to equivalence circuits described by models of network structures.[2] The results reveal the need to match the AC frequency with the characteristic time constant of a structure for optimum active and reactive power. For fuel-cell electrode conversion, the distribution of Pt and PtRu nanoparticles in a mesoporous carbon can vary in different synthetic protocol. Uniform Pt particle distribution can be achieved with a CPDP method [3] as shown in Fig. 2. Different electrocatalysts with varying Pt distribution can be compared for activity and utilization in methanol oxidation. The particles dispersed into a thicker shell are more stable and give better performance over time, as shown in voltage cycling and extended methanol oxidation.

Fig. 1. A hollow core mesoporous shell (HCMS) carbon structure with three varying geometrical parameters: pore diameter ($d_p$), shell thickness ($S$), and diameter of core ($C$).

Fig. 1. A hollow core mesoporous shell (HCMS) carbon structure with three varying geometrical parameters: pore diameter ($d_p$), shell thickness ($S$), and diameter of core ($C$).

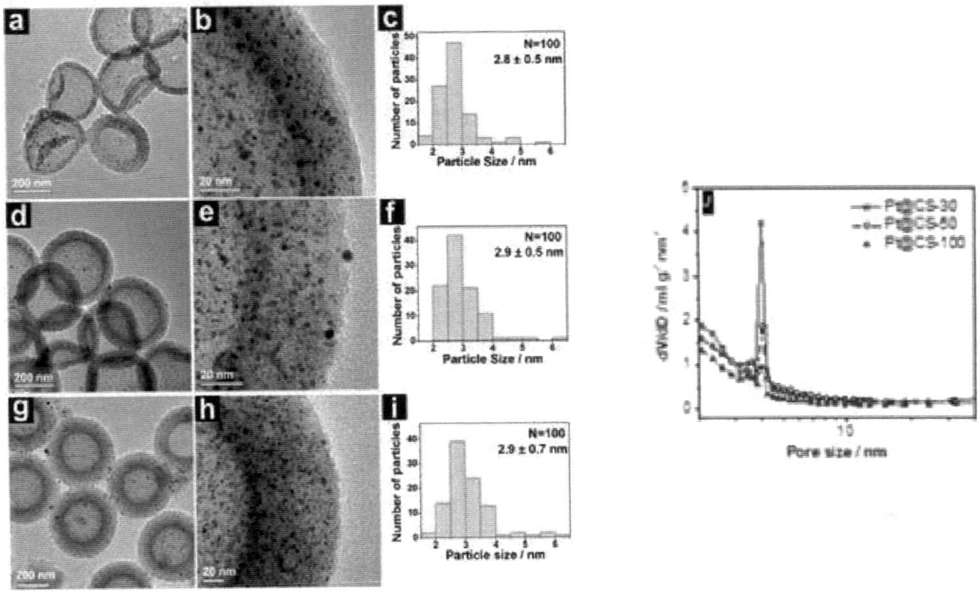


Fig. 2 TEM images of Pt nanoparticles loaded in mesoporous carbons of same hollow core diameter, $C$ = 330 nm but different Shell thickness (a)-(c) Pt@CS-30, $S$=30 nm ; (d)-(f) Pt@CS-50, $S$=50 nm ; (g)-(i) Pt@CS-100, $S$=100 nm; (j) pore size distribution of the porous structures loaded with Pt.

References:

1]. F. Li, M. Morris and K.Y. Chan*, ?Electrochemical Capacitance and Ionic Transport in the Mesoporous Shell of a Hierarchical Porous Core-Shell Carbon Structure?, *J. Mater. Chem.* 21 (2011) 8880-8886.

[2] Chunzhen Yang, Chi-Ying Vanessa Li, Fujun Li, Kwong-Yu Chan*, ?Complex Impedance with Transmission Line Model and Complex Capacitance Analysis of Ion Transport and Accumulation in Hierarchical Core-Shell Porous Carbons?, *J. Electrochem. Soc.* (2013) to appear.

[3] F. Li, K.Y. Chan*, and H. Yung, ?Carbonization over PFA-Protected Dispersed Platinum: An Effective Route to Synthesize High Performance Mesoporous-Carbon Supported Platinum?, *J. Mater. Chem.* 21 (2011) 12139-12144.

**(50e) High-Performance Electrodes for Hydroxide Exchange Membrane Fuel Cells**

Kaspar, R. B., University of Delaware
Letterio, M., University of Delaware
Jensen, K., University of Delaware
Gu, S., University of Delaware
Yan, Y., University of Delaware
Originally presented on: 11/4/2013 10:10:00 - 10:35:00

Hydroxide exchange membrane fuel cells (HEMFCs) are an attractive alternative to conventional proton exchange membrane fuel cells (PEMFCs) because they are compatible with non-precious-metal catalysts. However, HEMFC performance is lower than the PEMFC benchmark, in part because HEMFCs continue to borrow electrode architectures that were originally designed for PEMFCs. In PEMFCs, water is produced at the cathode, so hydrophobic gas diffusion layers (GDLs) are employed to control flooding. But hydrophobic GDLs may not be necessary for HEMFC cathodes, in which water is consumed instead of produced, reversing the direction of water transport in the cell. This work demonstrates that the performance and durability of membrane-electrode assemblies (MEAs) can be improved by tailoring electrode design to reflect the unique kinetics and water transport behavior of HEMFCs.

**(50f) Capacitive Mixing in Porous Electrodes for Energy Production**

Gabitto, J. F., PVAMU
Tsouris, C., Oak Ridge National Laboratory
Originally presented on: 11/4/2013 10:35:00 - 11:00:00

Energy obtained by salinity-gradients (blue energy) is a promising renewable energy source. Mixing of a high-concentration saline solution, such as seawater, with fresh water (river water) to produce a brackish solution dissipates more than 2.2 MJ of free energy per $m^3$ of fresh water. The charging-discharging cycle inside the pores of the solid electrodes is strongly influenced by the pore size distribution. Macroporous materials behave like standard plate capacitors, while mesoporous materials behave like cylindrical capacitors. The goal of this project was to optimize the recovery efficiency of free energy change by using tailored-design carbon electrodes. Morphology, pore size, and carbon electrode composition were chosen to optimize the charge-discharge steps of the reversible cycle. The physicochemical processes inside the electrode and the corresponding effects on the reversible energy cycle were investigated. A proposed energy cycle comprises a charging step, a discharging step and two switching steps that replace a saltwater solution inside the pores by a fresh water solution and vice versa. Equations derived using the volume averaging method have been employed to simulate the charging-discharging cycle in homogeneous porous materials of a macropore size distribution. Ionic transport by diffusion and electrical double layer expansion processes were considered to simulate the two alternating steps. Improved energy production was achieved by optimizing the duration of each step and variation of the operating variables.

**(74a) Process Systems Engineering Of Ionic Liquids**

Zhang, S., Institute of Process Engineering, Chinese Academy of Sciences
Zhang, X., Institute of Process Engineering, Chinese Academy of Sciences
Huang, Y., Beijing Key Laboratory of Ionic Liquids Clean Process
Xu, Y., Institute of Process Engineering, Chinese Academy of Sciences
Originally presented on: 11/4/2013 8:30:00 - 8:55:00

Ionic liquids (ILs) have been studied extensively as green solvents since the end of last century. The particular features and unique properties of ionic liquids make them have great potentials to be solvents, catalysts, electrolytes, coating materials and so on. Due to the serious and long-term pollutions caused by the conventional volatile organic solvents, replacement of conventional solvents by ILs is regarded with large benefits for sustainable and green chemistry [1]. Owe to their excellent solubilities in supercritical $CO_2$, the catalytic products are easily to be separated from reactants in an homogeneous catalysis system [2]. Lower volatility and high absorption capacity make ILs be promising candidates for carbon capture [3], Coating ILs on some solid materials like polymer membrane, or synthesis poly-ionic liquid membrane, can improve the selectivities of gas separation so as to reduce the separation energy consumption [4, 5]. It was reported that Ionic liquids showed good performance as a recyclable catalyst for cycloaddition of $CO_2$ [6] and for the degradation of waste polymer [7]. Recently, electroreduction of $CO_2$ which is a key component of artificial photosynthesis, has been proved that using ILs electrolyte can reduces $CO_2$ to CO at overpotentials below 0.2 volt which is just slightly above the minimum voltage of 1.33 volts [8]. In addition, physicochemical properties of ILs and modeling of ILs-based process have been studied extensively, which provides basic knowledge on applications of ILs [9-12]. These achievements show the promising industrial applications of ionic liquids, but also raise questions. The most arresting question is whether these processes can be commercialized in the near future; however, it is hard to find an answer.

Getting answerers to this question as early as possible is important and necessary, not only for researchers, but also for industries. We should not be expected to obtain completely final study results before we know the possibilities or the feasibilities of new IL-based processes. Hence, we propose the concept of 'Process Systems Engineering of Ionic Liquids' with the aim of investigating the advantages and drawbacks as well as the feasibilities of a new IL-based process, here we highlight the whole system rather than a single unit or a substance, and also take into account the whole life cycle rather than a short period. Thus, the key point of the concept is that it includes two dimensions, i.e., a spatial dimension and a temporal dimension. The former indicates the evaluated system including the whole facilities involved and the concept from molecular scale to process scale, while the latter means a whole life cycle should be considered, from resource extraction, IL production, IL utilization, IL recovery and degradation. The second point is the definition of the evaluation criteria, some are quantitative and some are qualitative, and also should be derived from the viewpoints of both academy and industry. Generally, these criteria include different aspects depending on the different requirements, such as energy efficiency, atom utilization efficiency, economic and environmental performance, and the potential risks. Moreover, how to build strict and robust models for simulation, prediction and optimization of basic properties, units, processes, costs, and environmental performance will still be a challenge.

With the methodology, some specific cases, such as carbon capture with IL, IL-based catalytic reaction, are studied. The results will be very useful for the development of new ionic liquid processes. It is also quite valuable to connect chemists with chemical engineers and promote a new idea from laboratory to br used in industries.

*Corresponding information: Email: sjzhang@home.ipe.ac.cn; Tel/Fax: 0086-10-82627080

**Acknowledgements**

We would like to acknowledge the support from Key Program of National Natural Science Foundation of China (No.21036007), the National Natural Science Foundation of China (No. 51274183), and Science and Technology Innovation "Cross and Cooperation Team" of Chinese Academy of Sciences (2013.1).

**Reference:**

(1) Rogers, R. D.; Seddon, K. R. Ionic liquids - Solvents of the future? *Science* **2003,** 302, (5646), 792-793.

(2) Cole-Hamilton, D. J. Homogeneous catalysis - new approaches to catalyst separation, recovery, and recycling. *Science* **2003,** 299, (5613), 1702-1706.

(3) Zhang, X. P.; Zhang, X. C.; Dong, H. F.; Zhao, Z. J.; Zhang, S. J.; Huang, Y. Carbon capture with ionic liquids: overview and progress. *Energy & Environmental Science* **2012,** 5, (5), 6668-6681.

(4) Li, P.; Paul, D. R.; Chung, T. S. High performance membranes based on ionic liquid polymers for CO2 separation from the flue gas. *Green Chemistry* **2012,** 14, (4), 1052-1063.

(5) Kim, D. H.; Baek, I. H.; Hong, S. U.; Lee, H. K. Study on immobilized liquid membrane using ionic liquid and PVDF hollow fiber as a support for CO2/N-2 separation. *Journal of Membrane Science* **2011,** 372, (1-2), 346-354.

(6) Sun, J.; Wang, J. Q.; Cheng, W. G.; Zhang, J. X.; Li, X. H.; Zhang, S. J.; She, Y. B. Chitosan functionalized ionic liquid as a recyclable biopolymer-supported catalyst for cycloaddition of CO2. *Green Chemistry* **2012,** 14, (3), 654-660.

(7) Wang, H.; Li, Z. X.; Liu, Y. Q.; Zhang, X. P.; Zhang, S. J. Degradation of poly(ethylene terephthalate) using ionic liquids. *Green Chemistry* **2009,** 11, (10), 1568-1575.

(8) Rosen, B. A.; Salehi-Khojin, A.; Thorson, M. R.; Zhu, W.; Whipple, D. T.; Kenis, P. J. A.; Masel, R. I. Ionic Liquid-Mediated Selective Conversion of CO2 to CO at Low Overpotentials. *Science* **2011,** 334, (6056), 643-644.

(9) Dong, K.; Zhang, S. J.; Wang, D. X.; Yao, X. Q. Hydrogen bonds in imidazolium ionic liquids. *Journal of Physical Chemistry A* **2006,** 110, (31), 9775-9782.

(10) Zhang, S. J.; Sun, N.; He, X. Z.; Lu, X. M.; Zhang, X. P. Physical properties of ionic liquids: Database

and evaluation. *Journal of Physical and Chemical Reference Data* **2006,** 35, (4), 1475-1517.

(11) Dong, K.; Song, Y. T.; Liu, X. M.; Cheng, W. G.; Yao, X. Q.; Zhang, S. J. Understanding Structures and Hydrogen Bonds of Ionic Liquids at the Electronic Level. *Journal of Physical Chemistry B* **2012,** 116, (3), 1007-1017.

(12) Tian, X. A.; Zhang, X. P.; Wei, L.; Zeng, S. J.; Huang, L.; Zhang, S. J. Multi-scale simulation of the 1,3-butadiene extraction separation process with an ionic liquid additive. *Green Chemistry* **2010,** 12, (7), 1263-1273.

**(74b) The Word Green, What Does It Mean?**

Ozokwelu, D. E., US Department of Energy

Originally presented on: 11/4/2013 8:55:00 - 9:20:00

The word Green is used rather so much these days and its definition means different things to different people, organizations or disciplines. To someone from the health industry, it is about using materials that are safe and non-toxic. For people working in measurements, monitoring and standards, it is about developing procedures and standards for monitoring particulates in the atmosphere, developing standards for biofuels, etc. For the energy sector it is about reducing energy and materials through energy efficiency, renewable energy, and developing better catalysts for efficient use of feedstock through process intensification. For the US department of commerce, it is about managing chemicals for export and having better international trade relations. This talk will examine the various uses of the word green by diverse individuals or groups or organizations from different government agencies, and private industries. The talk will close with an attempt to proffer a definition for the word green that will fit the meaning of the word green to all the different peoples, organizations and disciplines.

**(74c) Energy, Economy, Environment and Sustainable Development in the Middle East and North Africa**

Riazi, M. R., Kuwait University

Originally presented on: 11/4/2013 9:20:00 - 9:45:00

In this presentation we review the present state of energy demand and production in the Middle East and the world. Oil and gas reserves in the Middle East and North Africa and their role in providing energy demand by the world are examined. Role of oil and gas in the economy of the region and their impact on the global environment and carbon emission will be discussed.

Future of oil and gas reserves in the region and how long they will last along with existing coal reserves would be reviewed. Several models for the energy demand of the world by 2050 and the role of other sources of energy such as biofuel, nuclear, solar, hydro, wind, and wastes on sustainable energy supply will also be reviewed.

The role of population growth on the energy demand of the future and economy of the region will be discussed. The impacts of quality of crude oil and its products on the environment and changing fuel standards in the developed and developing countries and their effects on the market and industry as well as economy of the region will be reviewed.

The region is rich in oil and gas which can be converted into thousands of valuable products rather than burning as fuel. What factors affect the price of oil and gas in the future? Would that be economical to use oil and gas in refineries just to produce fuels or there are other options for more economical uses of these resources in the future? What would be the best option for the oil producing countries in the region? These issues and more will be presented in this speech.

**(74d) Advanced Zeolite Membrane-Enabled Air Dehumidification and Conditioning Technologies for Buildings**

LIU, W., Pacific Northwest National Laboratory
Originally presented on: 11/4/2013 9:45:00 - 10:10:00

Air conditioning in buildings consumes a large amount of electricity globally, which presents significant application opportunities for advanced water-selective membranes. The energy saving values are worth tens of billions $/year. Energy efficiency of the conventional vapor compression cooling is low in warm and humid climates due to water condensation. Greater than 50% energy efficiency may be achieved by removing moisture through a membrane without changing air temperature and pressure. The membrane dehumidifier also enables evaporative cooling with even higher efficiency and sustainable water. In addition, a membrane can be used to build a mass and heat transfer device for enthalpy recovery of ventilation, i.e., recovery of both latent and sensible energies.

The critical requirements for these applications are a low-cost membrane with high moisture permeance in addition to adequate $H_2O$/air selectivity. Therefore, thin-sheet zeolite membranes are developed in this work by deposition of the NaA-type zeolite membrane film on a robust, 50μm-thick porous metal support sheet. In this presentation, air dehumidification performances of the resulting zeolite membrane are compared to other existing and new membrane materials, such as polymeric membranes, ionic liquids and other immobilized solutions, and graphene oxide. The moisture permeance and $H_2O$/air selectivity measured under the same conditions differ dramatically among those membranes. More importantly, impacts of separation conditions (temperature, feed air humidity) on separation performances are highly dependent on specific membrane material, which can be explained by different separation mechanisms. Overall, the thin-sheet zeolite membrane exhibits the highest moisture permeance ($\sim 1\times10^{-5}$ mol $m^{-2}$ $Pa^{-1}s^{-1}$) among the membranes evaluated here and among the literature reports as well. The zeolite membrane also shows very desirable performance attributes for air dehumidification applications, such as increase of permeance and $H_2O$/air selectivity with air dehumidify and stable performances over a range of temperature. The thin-sheet zeolite membrane is promising for dramatic (>50%) enhancement of building air conditioning energy efficiency without any environmental emissions.

**(74e) Ionic Liquids In Biorefining and CO2 Capture**
Mikkola, J. P., Ã...bo Akademi University
Originally presented on: 11/4/2013 10:10:00 - 10:35:00

For centuries the Nordic countries have been relying on their natural resources, particularly forest, to make value-added products exported throughout the world. At the same time, this helped the societies to create wealth, job opportunities and social welfare. The global changes taking place today have gradually rendered these industries less competitive and profitable upon increasing competition when the consumption of such bulk products like paper and pulp have declined in the western word. The rise of the digital era, i.e. increasing use of electronic information and documentation, has meant that the volumes needed are in the decline in the developed world whereas the developing world still has room for growth in these products. Consequently, the industry has been forced to adapt and change, moving a lot of production capacity to there were the demand is. At the same time, increasing efforts are made to transform and trim the production in the original 'homeland' plants towards new, more innovative and more sophisticated products and, importantly, towards production of bio-based energy, transportation fuels, chemicals and commodities. This, however, is not easy since the industry is very conservative and the mindset is focused to maintaining the earlier business models.

Novel technologies based on the utilization of abundant and non-toxic carbon dioxide ($CO_2$) for the synthesis of materials and chemicals may significantly contribute towards the development of sustainable industrial practices. In nature, myriad of substances are photo-synthesized from $CO_2$ and water and consumed by humans. The use of $CO_2$ in the synthesis of 'switchable' ionic liquids (SILs) constitutes an environmentally attractive alternative to hazardous and toxic reagents thus resulting in waste minimization. 'Switchable' ionic liquids are compounds which can be switched from molecular compounds to ionic liquid by bubbling an acid gas such as $CO_2$ or $SO_2$ through a liquid mixture of precursors. The formed ionic liquid can typically be switched back to molecular building blocks – and the acid gas released - by bubbling with an inert gas (e.g. $N_2$), rising the temperature to above the thermal decomposition temperature of the SIL or by vacuum distillation at sufficiently high temperatures.

As another example, new materials such as polymeric ionic liquids were utilized to selectively remove carbon dioxide. Also, the performance of 'classical', 'switchable' and polymeric ionic liquids was compared to that of established, alkanol amine based solutions. In essence, our groups have synthesized and successfully demonstrated the use of new families of various kinds of ionic liquids including BILs and SILs in applications such as fractionation of lignocellulose, cellulose dissolution, purification of biogas and $CO_2$ capture. SILs are also attractive in terms of synthesis practices since no special organic chemistry laboratory infrastructure is needed. Moreover, they are surprisingly tolerant towards water and do not lose their ability to dissolve even humid biomass.

In this talk an overview of the technological efforts towards future bio-based economies and industries will be given in light of selected research projects currently in progress.

## (74f) Process Intensification and Coupling Effects for Efficient Conversion of Biomass Based On Ionic Liquids

Li, X.,
Long, J.,
Ma, H.,
Originally presented on: 11/4/2013 10:35:00 - 11:00:00

It is well known that great progress achieved in material civilization for human beings in the last century was mainly on the expense of non-renewable global resources, which has brought lots of realistic problems such as energy crisis, environmental pollution and climate change. Hence, the exploitation and development of novel and cleaner technologies for efficient utilization of carbon neutral biomass have been recognized as one of the most predominated approaches for sustainable development of society. Therefore, efficient and direct transformation of abundant and renewable biomass to value-added bioenergy and chemicals is of great significance, which is in good agreement with the concept of low-carbon economy.

In this presentation, a series of novel and functionalized ionic liquids were reasonably designed and synthesized for efficient conversion of biomass through the investigation of the structure-active relationship between ionic liquids and the dissociation behavior of biomass. Furthermore, process intensification and coupling effects for efficient conversion of biomass has been realized by constructing reaction-separation systems of ligincellulose dissolution-*in situ* catalysis-product extraction/reaction and in subercritical fluids. Moreover, novel chloro-functionalized ionic liquids as effective catalysts for the conversion of cellulose to Levulinic acid will be reported.

Reference

1. J X Long, X H Li, B Guo, L F Wang, N Zhang. *Catalysis Today*, 2013, 200(1): 99-105.

2. J X Long, X H Li, B Guo, FR Wang, YH Yu, LF Wang. *Green Chemistry*, 2012, 14(7): 1935-1941.

3. J X Long, X H Li, LF Wang. *Science China B: Chemistry*, 2012, 55(8): 1500-1508 .

4. J X Long, B Guo, J J Teng, Y Y Yu, L F Wang, X H Li*. *Bioresource Technology*, 2011, 102(21): 10114-10123.

5. J X Long, B Guo, X H Li, F R Wang, Y H Yu, L F Wang. *Acta Physico-Chimica Sinica*, 2011, 27(5): 995-999 .

6. J X Long, B Guo, X H Li, Y B Jiang, S C Tsang, L F Wang, K M K Yu. *Green Chemistry*, 2011, 13(9): 2334-2338.

**(118a) Sustainability Progress**
Abraham, M. A., Youngstown State University
Originally presented on: 11/4/2013 12:30:00 - 13:07:00

Sustainability concepts are critically important in the chemical industry. From a practical perspective, the implementation of sustainability throughout the process industries is increasingly important to our society, and thus to the success of the chemical industry. But sustainability is more than simply green chemistry and engineering, as it involves life cycle and societal implications. Perhaps more so than elsewhere, the production of energy and its use illustrates the importance of sustainability in chemical processes. Interdisciplinary concepts have become the hallmark of sustainable chemical design, synthesis, process and use, and influence sustainable development of existing or new chemicals and materials using a holistic approach. This presentation will describe recent activities in the application of sustainability and sustainable energy as they have been presented over the past few years in the AIChE quarterly *Environmental Progress and Sustainable Energy*.

**(118b) Sustainable Biofuels: A Case Of Ethics, Politics, Economics and (I almost forgot) Technology**

Sheehan, J. J., NREL

Originally presented on: 11/4/2013 13:07:00 - 13:45:00

The sustainability of biofuels is inextricably linked to the broader challenges of sustainable agriculture and sustainable development. And both these challenges are deeply rooted in political and ethical questions that we have yet to address in a clear and coherent manner. The debate about biofuels' indirect effect on global land use change is the latest and clearest connection between these two problems. Indirect land use change, simply put, considers the problems and possibilities of agricultural expansion and subsequent land clearing associated with increased demands on agricultural land from biofuels.

This land use debate is framed using the "objective" paradigm of the scientific experimenter, which I contend has led us to waste precious research and analysis resources on the wrong question. In this paradigm, biofuel's sustainability rests on a starting assumption that all other aspects of our global agriculture system remain unchanged (the experimental paradigm of ceteris paribus). This is tantamount to asking if biofuels, when added to an already unsustainable agriculture system, has any chance of being sustainable. The answer is obvious: No. But we have spent million of dollars and hours seeking an analytical answer to this question rather than focusing on the questions that really matter. First and foremost, what does it mean for biofuels, agriculture and development writ large to be sustainable? This is as much an ethical question as it is a technical and economic one. In my talk, I will highlight some of the ethical aspects of sustainable development to illustrate how they influence our conclusions about biofuels. Second, what must be done to ensure that global agriculture gets on a path to sustainability? Here, I will discuss a few aspects of modern day global agriculture that offer powerful levers for change. Third, can and should biofuels play a role in sustainable development? The question of "can" largely depends on the answers to the previous question. The question of "should" is a practical and ethical question the answer to which (I will argue) is yes.

**(118c) NSF and Support Of Sustainable Engineering Research and Education**
Hamilton, B., National Science Foundation
Originally presented on: 11/4/2013 13:45:00 - 14:22:00

An overview of NSF support of sustainable engineering research and education will be presented. Funding opportunities will be highlighted, and brief examples of funded research and education activities will be given. Comments will be made on some potential future directions for research on sustainable engineering. NSF has a wide variety of sustainable engineering research funding opportunities available. At NSF, funding opportunities can be grouped into two general categories, "standing programs" and "solicitations." Standing programs accept proposals on a regular basis, usually at least once per year, during a date window for each program (see the webpages for specific programs on NSF's website,www.nsf.gov ). For engineering research on sustainability, NSF's Engineering Directorate (ENG) has established two standing programs that have the word "sustainability" in their program titles. There two programs are the "Environmental Sustainability" program and the "Energy for Sustainability" program. Both of these programs have webpages on NSF's website that describe the programs and how to apply to them for research grants. On each of these program pages, links are embedded that list all the active grants funded by each program, with a description of each grant. In ENG, other programs also accept proposals related to sustainability. For example, the Process and Reaction Engineering program accepts proposals related to the sustainability of chemical processing and bioprocessing. The Civil Infrastructure Systems program will consider proposals on infrastructure sustainability and resilience. The Manufacturing Enterprise Systems program will consider proposals on supply chain sustainability. These are just a few example opportunities in the standing program category. Grants in this category are typically $300K with a duration of 3 years, or, for CAREER grants (Assistant Professors only), $400K with a duration of 5 years. "Solicitations" are special calls-for-proposals, and may be issued only once, or at most, just a few times. For sustainability research, NSF recently established a Foundation-wide investment area, Science, Engineering and Education for Sustainability (SEES). SEES is offering a good number of solicitations, providing diverse opportunities for research proposals. Among these opportunities is the Water Sustainability and Climate solicitation (WSC), with new grant opportunities up to $4 million each. Another solicitation is entitled "Sustainability Research Networks (SRN)." SRN awards are up to $12 million each. One SRN competition has been held so far and resulted in two SRN awards. One of these SRN awards is on the environmental and social impacts of fracking. The other is on climate change mitigation. A second SRN competition is being planned, and should be announced within a few months. Additionally, about 25 grants have been made under SEES to establish Research Coordination Networks (RCN-SEES grants). These are just a few of the activities under SEES, and more information on SEES is available on NSF's website.

**(118d) Sustainable Development In The Face Of Water Stress**

Reible, D. D., Texas tech University

Originally presented on: 11/4/2013 14:22:00 - 15:00:00

Recent droughts focused attention on the fact that our future is endangered in many parts of the world. Our traditional sources of water are finite. Yet our population will grow and with it our demands for water to support agriculture, economic development and basic human needs. This can only make our economy more sensitive to drought. Moreover, climate trends suggest that water scarcity may become more common and more intense. Without solutions, our growth will ultimately be limited by water scarcity and droughts will have an impact more damaging than any traditional economic recession.

Certainly part of the solution is conservation and demand reduction by all users including industry. It is likely, however, that conservation and demand reduction can at best offset population and economic growth. It is likely that only the implementation of new technologies to advance reuse and recycling and to upgrade poor quality water resources will ensure that water does not ultimately limit economic development. Advanced technologies will be necessary to enable wastes to be treated to an extent to allow greater water reuse and to upgrade poor water sources such as saline groundwaters and produced water. Implementation of these technologies in a manner that will not be equally damaging to the economy, however, will require significant advances in their energy efficiency, reliability and effectiveness.

The presentation will explore these topics, making the case for future needs and discuss some of the challenges and opportunities for technology development to address those needs. The presentation will focus on the application of these approaches to minimize the fresh water requirements of hydraulic fracturing for the extraction of oil and gas from shale.

**(146a) Development of Composite Hollow Fiber Membranes for Aqueous and Non-Aqueous Nanofiltration**

Sun, S. P., National University of Singapore
Chan, S. Y., National University of Singapore
Chung, T. S., National University of Singapore
Originally presented on: 11/4/2013 12:30:00 - 12:55:00

Nanofiltration (NF) has been growing rapidly in pharmaceutical separation, wastewater treatment, seawater desalination and many other applications. This presentation focuses on the new developments of composite NF hollow fiber membranes from both co-extrusion and interfacial polymerization (IP). Firstly, a highly water permeable dual-layer NF hollow fiber membrane fabricated by the simultaneous co-extrusion of Torlon® polyamide-imide (PAI) and cellulose acetate (CA) polymer solutions through a triple-orifice spinneret in a dry-jet wet phase inversion process. As compared to conventional membranes, the dual-layer hollow fiber membranes developed here have a unique structure which consists of (1) controllable thin selective outer layer; (2) innovatively designed interface, which is finely connected between outer and inner layers, does not contribute significant transport resistance while still maintains good mechanical strength; (3) fully porous substructure with reduced transport resistance. The cross-section struction and the nanopores of dual-layer hollow fiber membranes were molecularly designed by controlling the phase inversion process with the aid of various non-solvent additives into the polymer solutions. The newly developed NF dual-layer hollow fiber membrane with methanol as additive has a relatively high pure water permeability of 11.93 $l\ m^{-2}bar^{-1}h^{-1}$ with a mean effective pore radius of 0.63 nm and a more than 96.5% rejection of $Na_2SO_4$.

Secondly, a novel thin-film composite NF membrane was fabricated by IP of hyperbranched polyethyleneimine (HPEI) and isophthaloyl chloride (IPC) on a dual-layer hollow fiber membrane. The substrate possesses a unique cross-section structure comprising a layer full of macrovoids located in the middle and sandwiched by two thin spongy-like layers. Such a sandwich-like sublayer structure provides minimal transport resistance and sufficient mechanical strength for water permeation under high pressures. After IP, the NF membrane possesses a negatively charged substrate and a positively charged selective layer with a mean effective pore radius of 0.36 nm, molecular weight cut off of 500 Da, and pure water permeability of 4.9 $lm^{-2}bar^{-1}h^{-1}$. Due to this double-repulsion effect, together with the steric-hindrance and the solute electro-neutrality effects, by adjusting the pH of cephalexin aqueous solution to modify the ionization states of this zwitterionic molecule, the NF membrane shows high rejections over a wide pH range. Furthermore, the latest preliminary data shows the membrane is very stable in harsh organic solvents such as dimethyl sulfoxide. Therefore, the NF membrane may potentially be useful to reduce waste, recycle valuable pharmaceuticals and reuse solvents for the pharmaceutical industry.

**(146b) Self-Heat Recuperative Seawater Desalination Process Using Fluidized Bed**

Mizuno, H., The University of Tokyo
Kansha, Y., The University of Tokyo
Ishizuka, M., The University of Tokyo
Tsutsumi, A., The University of Tokyo
Originally presented on: 11/4/2013 12:55:00 - 13:20:00

Recently, it is presumed that there will be a serious shortage of water in several years later due to the high growth rate of population and industrial development. To alleviate the water shortage, more attention has been paid to desalination technology. One of the typical desalination processes is a multi-stage flash desalination. This process is categorized as thermal desalination process because it produces pure water by evaporation. The thermal process has the following two major advantages; seawater quality does not seriously affect the operation condition and product water salinity is low enough for industrial usage. However, this process consumes a lot of energy, thus the remarkable energy saving technology of thermal process is in great demand. So far, to reduce the energy consumption, we proposed a novel thermal desalination based on self-heat recuperation, which is a method for energy saving. By a process simulator, it was revealed that the proposed desalination process energy consumption is reduced to 1/4 of conventional thermal desalination process. Furthermore, it was elucidated that the specific energy consumption decreased as the recovery ratio (= the product water flow rate / the feed water flow rate) increased. However, in the thermal desalination process with high recovery ratio, a problem of scale deposition, which leads to a reduction of heat transfer efficiency in heat exchanger occurs.

In this research we introduce self-heat recuperative seawater desalination process with fluidized bed for solving the problem. In the desalination process, the fluidized bed is employed as an evaporator. It might be seen that this process has the following advantages; good heat transfer properties and less scale deposition because of water evaporation taking place on heated particles in fluidized bed evaporator.

In addition, we investigated the influence on the specific energy consumption and feasibility to the society of the proposed desalination process.

**(146c) Characteristics of Water and Salt Transport, Free Volume and Their Relationship With the Functional Groups of Novel Cellulose Esters**

Ong, R. C., National University of Singapore
Helmer, B. J., Eastman Chemical Company
de Wit, J. S., Eastman Chemical Company
Chung, T. S., National University of Singapore
Originally presented on: 11/4/2013 13:20:00 - 13:45:00

Fundamental transport properties including salt and water partition coefficients, permeability and diffusivity of various newly synthesized cellulose esters were evaluated in this study in order to investigate the relationship among them as a function of molecular structure. Dense flat cellulose ester films were prepared and studied so that the influence of membrane fabrication, morphology and processing history on transport properties can be minimized. Annihilation lifetime spectroscopy (PALS) was employed to characterize the free volume of these cellulose esters. It was found that the transport and free volume properties can be correlated with the functional groups and their content in cellulose esters. A high hydroxyl content leads to a high hydrophilicity but results in a high water and salt partition coefficient. Hydrophilic cellulose esters suffer from a low water/salt selectivity. On the other hand, an increase in bulky and hydrophobic propionyl or butyryl groups generally enhances cellulose esters' selectivity but brings down the permeability. However, a very high content of bulky functional group causes a drop in selectivity due to poor chain packing and the enlarged free volume as confirmed by PALS. Experimental results suggest that cellulose esters with moderate content of both hydroxyl and bulky side groups have the best diffusivity selectivity and may be suitable as membrane materials for salty water separation. An upper bound relationship between solubility selectivity and water partition coefficient was also observed in these cellulose esters as suggested in previous literatures for other polymers.

**(146d) Bioremediation of Cu(II) Contaminated Aqueous Solution Using Acinetobacter Sp**

Majumder, S., Birla Institute of Technology & Science, Pilani
Gangadhar, G.,
Raghuvanshi, S., Birla Institute of Technology & Science, Pilani
Gupta, S., Birla Institute of Technology & Science, Pilani
Originally presented on: 11/4/2013 13:45:00 - 14:10:00

Bioremediation of metal wastes using enriched mixed consortium of microbial species is a very promising and viable alternative in environmental biotechnology. Present study deals with the removal of divalent copper [Cu(II)] compound from aqueous solution using *Acinetobactor sp.* (GenBank Accession Number: HQ538658.1), an indigenous bacterial strain which is isolated from enriched mixed culture. Enrichment of mixed culture is carried for a period of 6 days. At the end of the enrichment period, the biomass concentration obtained is 0.0187 g/L of solution. The effect of different operating conditions such as contact time, initial pH and amount of nutrient dosage on bioremediation of Cu(II) are studied in a batch mode. The effect of contact time is studied for an initial Cu(II) concentration ranging from 60-150 mg/L. Increase in initial Cu(II) concentration from 60 mg/L to 150 mg/L shows better removal of Cu(II). The effect of initial pH is also studied by varying the solution pH in the range of 2 - 12. The maximum biomass growth obtained is 0.112 g/L at pH = 7.5 for initial Cu(II) concentration of 150 mg/L. The maximum percentage removal of Cu(II) is obtained for the initial Cu(II) concentration of 60 mg/L at the solution pH value of 7.5. Percentage removal of Cu(II) value is also compared with reported value of percentage removal of Cu(II). This study also shows that the microbial growth for Cu(II) removal is significantly affected by nutrient dosage. The optimum biomass growth obtained is 0.0248 g/L at double of the standard composition of MSM.

***Keywords:*** *Acinetobactor sp., Divalent copper; Bioremediation; Acclimated mixed culture, MSM.*

**(146e) Calcium Ions in Gypsum (CaSO4Â·2H2O) Scaling in Forward Osmosis Processes and a Novel Design Strategy to Fabricate Polybenzimidazole Anti-Scaling Hollow Fiber Membranes**

Chen, S. C.,
Su, J., National University of Singapore
Fu, F. J.,
Mi, B., The George Washington University
Chung, T. S., National University of Singapore
Originally presented on: 11/4/2013 14:10:00 - 14:35:00

**Calcium ions in gypsum ($CaSO_4$Â°Â¤$2H_2O$) scaling in forward osmosis processes and a novel design strategy to fabricate polybenzimidazole anti-scaling hollow fiber membranes**

Si Cong Chen[a,b], Jincai Su[b], Feng-Jiang Fu[b] , Baoxia Mi[c], Tai-Shung Chung[a,b,*]

[a] NUS Graduate School for Integrative Sciences & Engineering (NGS), National University of Singapore, 28 Medical Drive, Singapore 117456, Singapore

[b] Department of Chemical and Biomolecular Engineering, National University of Singapore, 4 Engineering Drive 4, Singapore 117576, Singapore

[c] Department of Civil and Environmental Engineering, University of Maryland, College Park, MD 20742, USA

* Corresponding author: chencts@nus.edu.sg

Abstract

Gypsum ($CaSO_4$Â°Â¤$2H_2O$) scaling phenomena have been studied on hollow fiber membranes with different physicochemical properties in forward osmosis processes. Three types of hollow fiber membranes were demonstrated i.e., (1) cellulose acetate (CA), (2) polybenzimidazole (PBI)-polyethersulfone (PES) and (3) PBI-polyacrylonitrile (PAN) - polyhedral oligomeric silsesquioxane (POSS). For the first time, we have found that calcium ions not only dominate but also take a leading role on gypsum ($CaSO_4$Â°Â¤$2H_2O$) scaling through ionic interactions with the membrane surface. A high level of calcium ions in scaling solutions can initiate gypsum crystallization more effectively than a high level of sulfate ions. Based on this finding, for the first time, a novel strategy to molecularly engineer PBI hollow fiber membranes with superior anti-scaling properties has been proposed and demonstrated. As shown in Figure 1c, the newly designed PBI-PAN (POSS) membranes have a positively charged and smooth membrane surface at the nanoscale level but possess a ridge and valley structure at the micrometer level. It shows slightly positive charge at a low pH and exhibits minimal scaling of only 5% flux reduction in contrast to CA membranes with 63% flux reduction. The adhesion force measured by atomic force microscopy (AFM) confirms a strong repulsive force between gypsum particles and PBI (Figure 1b) at pH values lower than the isoelectric point of 5.7. The repulsion force leads to the anti-fouling properties, while the novel ridge and valley morphology in the micrometer scale facilitates surface turbulence flow and effectively remove foulants.

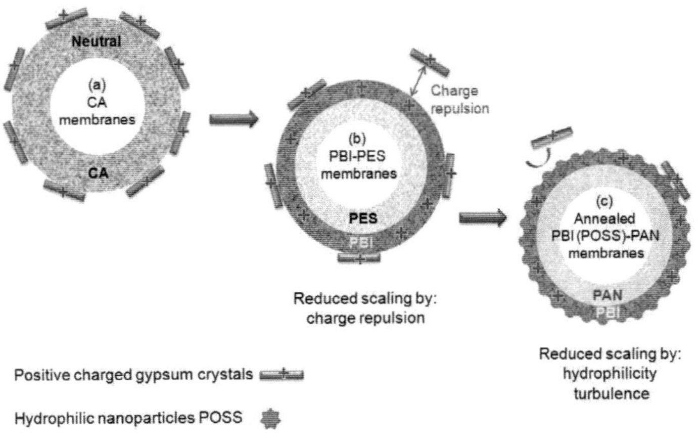


Figure 1: Fouling mechanisms of gypsum scaling on membranes

**(146f) New Insights Into Hydrophobic/Hydrophilic Composite Hollow Fibers for Direct Contact Membrane Distillation**

Zhu, J. M., Central South University
Jiang, L. Y., Central South University
Originally presented on: 11/4/2013 14:35:00 - 15:00:00

Hydrophobic/hydrophilic composite membrane is a new design of membrane in terms of structures and materials for DCMD. With dry-wet phase inversion process, a novel type of hydrophobic/hydrophilic hollow fiber membranes was fabricated in this research. The outer layer was made from polyvinylidene fluoride (PVDF), while the inner layer PVDF and polyvinyl alcohol (PVA) blend. Effects of dope composition, PVA/PVDF blending ratio and coagulants on the morphological, physiochemical and mechanical characteristics of the membranes thus generated were investigated. The pore size and porosity in respective layer of the dual-layer membranes were determined mainly by data obtained from SEM and corresponding software processing. The DCMD performance observed in experiments was compared with that obtained by mass and heat transport modeling of both single-layer and dual-layer DCMD membrane. Analysis relating the membrane structural features and the above comparison revealed that the inner layer characteristics in terms of both material hydrophilicity and pore size should be optimized to give a better membrane performance. The maximum flux in DCMD obtained was around $24.4 kgm^{-2}h^{-1}$ at 65℃.

**(182a) Economic and Engineering Assessment of Combined Heat and Power Fuel Cell Systems Installed in Commercial Buildings**

Brooks, K., Pacific Northwest National Laboratory
Pilli, S.,
Originally presented on: 11/4/2013 15:15:00 - 15:37:00

PNNL support of the DOE's Fuel Cells Technology Office is to identify near term niche markets where fuel cells have potential, work with the DOE and stakeholders to develop activities in those areas, analyze the business case, and present the results to the community. Accelerating fuel cell use in these key early markets will create jobs in an industry that needs high volume purchases to ramp up production, decrease costs, support commercialization, and enable a domestic supplier base. PNNL has contracted ClearEdge Power to demonstrate fifteen 5 kW combined heat and power fuel cells in a wide variety of business in California and Oregon. These fuel cells are being remotely monitored for such things as electrical and thermal energy, overall efficiency, and availability. The gathered information will be analyzed, to provide "real-world" data from units "in the customer's hands" to validate performance, durability and reliability; installation, operations, and maintenance costs; and identify remaining barriers to widespread commercialization. As part of this assessment, a business case has been developed estimating of the size of the market and its growth potential, identifying possible niche markets, and comparing the CHP FCS with its alternatives in terms of economics, engineering and environmental impact. The results of this assessment will be presented.

**(182b) Systematic Approach for Energy Efficient Eco-Industrial Parks**

Noureldin, M. B., Saudi Aramco
Qahtani, S. A., Saudi Aramco
Owaidah, M., Saudi Aramco
Nutaifi, A., Saudi Aramco
Originally presented on: 11/4/2013 15:37:00 - 15:59:00

Sustainability since its original conceptualization is considered to have four dimensions; social; economic; environmental and institutional. It is simply the improved management of natural resources within a business establishment to provide economic and social benefits to the business and its surroundings. Eco-industrial parks (EIP) can serve a significant role in realizing economic, environmental and social benefits both to individual plants and to network of companies. Even though EIP are often publicized as a mean of reducing environmental damage through reduced waste, they can have number of benefits at different levels. Currently the major driving factor for the creation of most of it is to attain financial gain.

Industrial ecology concept may promote a new path of local development through the transition from industrial clusters to EIPs via exploiting common features of both models. Implementing Eco-Industrial Parks' (EIPs) principles in an existing industrial cluster or in the planning of new ones, represents a significant opportunity for its revitalization. IEP could exploit synergies from industrial clusters and non-industrial activities to easily create new production models in which the economic and environmental dimensions are symbiotic.

The well planning of new EIP or transformation of conventional industrial complexes to a one can brings significant value to energy efficiency and energy-based GHG emissions reduction. This task is a huge multi-variable multi-dimensional optimization problem in which the total eco-industrial park network depends on as small as stream condition and as big as the whole park functionality. The customized integration among many industrial plants and non-industrial ones in adjacent geographical locations can bring in more degrees of freedom to the "waste energy recovery" optimization problem and consequently presents new horizon to the radical energy-based GHG emissions reduction to levels never thought of before. Fortunately, many of the reasons which were hindering the application of inter-systems integration are nowadays addressed in cost-effective ways. However, to date EIPs key success has been a sequence of independent economically driven actions. Such evolutionary pattern may not be easily transferred to others. A holistic approach in addressing the problem using systematic techniques in the implementing of necessary modifications in either the brownfields or in the planning of Greenfields can bring huge value to energy resources conservation and GHG emissions reduction and is desperately needed today. It is very difficult with the current methods and tools to magically manufacture EIPs to work from scratch. First, there should be the basic ingredients in place, namely the desire of plants to actively participate and second the correct EIP membership and structure of firms. These basic ingredients can then be enhanced and improved upon, with correct support structure in place. The key to developing a successful EIP is to determine the best connections among different industrial plants and its surrounding communities in an EIP. Specifically it is the material and energy flows relationship among the different members in the alliance of plants which permit establishing optimal linkage to form a fruitful inter-dynamic structure. If such structure does not exist a successful EIP cannot be realized.

This paper addresses the energy efficiency of EIP via new systematic approach. It is different in concept than the current approach in which only the "transfer of waste heat" from one firm to a nearby one is conducted after exhausting all the possibilities to optimize the stand-alone plant in the firm. The total site targeting based on the pinch technology, which is the most widely used to date, starts with intra-systems integration which means again that only the waste heat of one member in the EIP is shared with other parks' members. Certainly, such level of sharing philosophy is not the ideal proactive form of true cooperation.

This paper introduces new method that achieves ambitious levels of energy efficiency and GHG emissions reduction via identifying best matchings' among EIP members; generating technically viable energy efficient alternatives; and defining the best generation and allocation of energy utilities systems. It also presents two new mathematical programming and constraints logic propagation tools to systematically find the best and the second best and so on matchings' among several plants for optimal eco-industrial park membership forms under all possible combinations of plants-specific design and operations modifications. The paper also introduces new graphical tool that enables the EIP designer to gain insights and create initial solutions for optimal energy loads for exchange among the members to best enhance the energy efficiency and GHG emissions of the whole park ("One for all and all for one philosophy"). The advantages of the new method and tools are illustrated and demonstrated by two case studies.

**(182c) The Potential Of Osmotic Membrane Dehumidification**

Kesten, A. S., Nanocap Technologies LLC
McCutcheon, J. R., University of Connecticut
Girelli, A.,
Blechner, J. N., Nanocap Technologies LLC
Originally presented on: 11/4/2013 15:59:00 - 16:21:00

THE POTENTIAL OF OSMOTIC MEMBRANE DEHUMIDIFICATION

Arthur Kesten, Jeffrey McCutcheon, Ariel Girelli and Jack Blechner

An osmotic membrane dehumidifier can use a flexible, semi-permeable membrane to facilitate capillary condensation of water vapor and the transport of condensed water through the membrane into a salt solution by osmosis. Here a humid gas stream is brought into contact with a semi-permeable membrane, which separates the gas stream from an osmotic (e.g., salt) solution. Some of the pores of the membrane are small enough to permit capillary condensation. Liquid formed within these pores can connect with liquid formed in adjacent pores, collectively forming continuous paths of liquid. These ?liquid bridges' extend across the thickness of the semi-permeable membrane and provide paths by which water can travel across the membrane. Because the membrane is so thin, water concentration gradients across the membrane can be large. This can provide a large driving force for water transport between the humid air and the osmotic fluid. The flexibility of the polymeric membrane allows for considerable design flexibility that enhances the potential for retrofit with any cooling system. An illustration of this two-step spontaneous process is given below.

Liquid desiccants are particularly effective as osmotic agents because water entering the desiccant solution is bound to the salt as water of hydration. This enhances the water concentration gradient across the membrane.

Laboratory testing of membranes and draw solutions under different environmental conditions is conducted using a cell comprised of two halves separated by a membrane. On the top half, the humid air is passed above the membrane. The draw solution is pumped through the bottom half of the cell; the draw solution is contained in a reservoir that is placed on a digital scale that measures mass to a hundredth of a gram. Humid air contacting the membrane condenses by capillary condensation in the pores of the membrane and the condensed water is drawn by osmosis into the draw solution. Mass changes versus time are recorded with the digital scale that is capable of measuring a maximum of 4,000 grams.

Osmotic dehumidification begins with condensation of water molecules, followed by the osmotic draw of the liquid water out of the pores into a draw solution. If the rate-limiting step is the osmotic de-swelling of the membrane, the highest osmotic pressure should be used to maximize flux. However, if capillary condensation is the limiting step, a threshold is reached beyond which increased osmotic pressure does not increase flux. To understand these limits, we tested three concentrations of magnesium chloride with a cellulose acetate membrane under identical conditions. The water flux varied directly with concentration suggesting that the limiting step for this membrane was the osmotic removal of liquid water as opposed to capillary condensation.

A separate experiment was constructed to compare the effectiveness of no membrane to our capillary condensation/osmotic dehydration system. Having no membrane in the system is the equivalent of exposing liquid desiccant directly to a humid air stream. Measured water removal rates for that system were less than one fifth of the rates for the membrane/draw solution tested here. And, of course, with no membrane between the humid air and the desiccant, there is always the potential for entraining desiccant in the air stream.

Cooling of the osmotic solution results in more effective dehumidification. Lower water vapor pressure leads to capillary condensation at lower relative humidity. For modest levels of cooling, the relative humidity exiting the Nanocap process will be around 50%.

Scale up of the process to larger surface areas of membrane can be accomplished with plate and frame systems, spiral wound devices and bundles of cylindrical fibers.Â Results with a plate and frame system demonstrate the effectiveness of a mathematical model as a basis for design of a larger system.

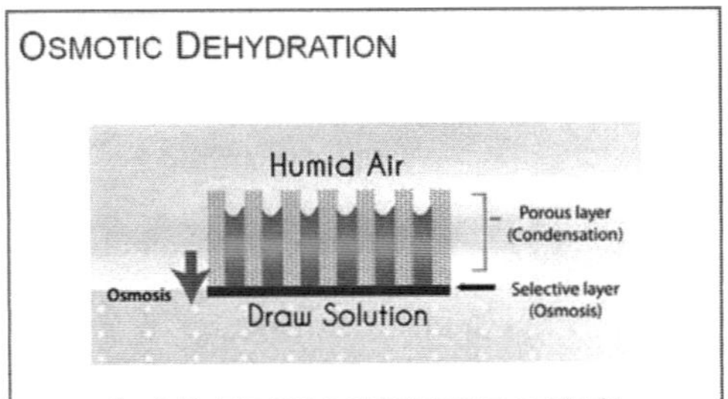


Osmotic dehydration across an anisotropic membrane consisting of a selective layer that is supported by a porous layer. Capillary condensation allows for osmosis to occur across the selective layer of the membrane into a concentrated salt or osmotic solution.

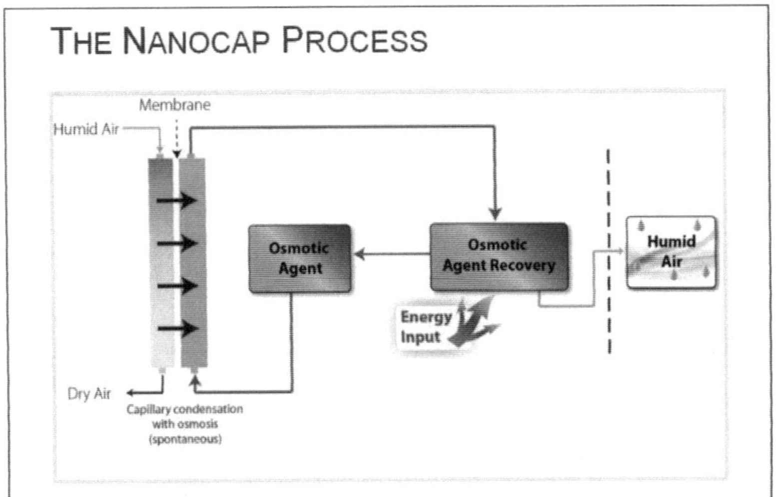


Energy used to regenerate the osmotic agent is intended to be a solar or waste heat source. Regeneration of the osmotic agent can be continuous or intermittent; the optimum time to regenerate occurs when solar or reject heat is available.

**(182d) Improvement of Distillation Column Efficiency By Integration With Organic Rankine Power Generation Cycle**

Sladkovskiy, D. A., St.Petersburg State Institute of Technology (technical university)
Originally presented on: 11/4/2013 16:21:00 - 16:43:00

Distillation as main separation technology consumes greater than 40% of the energy used by the chemical and petroleum process industries but the energy efficiency of a distillation column is low. It is well-known that significant energy savings can be achieved if complex column configurations and heat integration are used, although these improvements withdraw the attractive advantages of conventional distillation column such as flexibility, low capital investment and low operational risk.

This theoretical research focuses on improvement of distillation column vapor generation procedure by expansion turbine installation.

The liquid stream from the column bottom is pressurized by a pump and the high pressure liquid is vaporized in the reboiler. The resultant high pressure vapor stream drives the expansion turbine, which is coupled to the electricity generator or another rotating equipments. The turbine exhaust vapor enters the column bottom.

Such distillation column with power generation cycle (DCPGC) is kind of combination of conventional distillation column and organic Rankine cycle (ORC) power plant. The ORC is a cost efficient and proven method of converting low temperature waste heat to mechanical and/or electrical energy.

Such chemical components and its mixtures as propane, i-butane, n-butane, i-pentane et al. which occurs in distillation columns bottoms are known to be utilized or considered as ORC working fluids. So DCPGC is seem to be practically feasible.

DCPGC reboiler duty is about 10-20% greater than the conventional column reboiler duty but almost whole of extra heat transforms to electricity power. The theoretical DCPGC electricity generation efficiency is more than 90% (the heat required for the distillation remains the same as in conventional column and is not included in the efficiency calculation).

Only thermodynamic properties and thermal stability of column's bottom product impact on DCPGC feasibility, but not the temperature difference between top and bottom of the column. This is an advantage of suggested technology to compare with heat integrated column with heat pumps where temperature difference is quite important.

This theoretical research describes practical and economical feasibility of DCPGC for separating propane/propylene, i-butane/n-butane, i-pentane/n-pentane, butane-pentane, benzene/toluene, xylene mixtures. Crude oil and naphtha fractionation are also investigated. For instance DCPGC for i-butane/n-butane binary mixtures separation with 25 tonne/h feed flowrate can produce 1 MW of electricity power.

**(182e) High Surface Area Zeolite Membrane for Efficient Bio-Fuel Production**
LIU, W., Pacific Northwest National Laboratory
Originally presented on: 11/4/2013 16:43:00 - 17:05:00

Distillation and adsorption has been commonly used for respective enrichment and polishing of ethanol fuels in today's ethanol plants. In this presentation, development of thin zeolite/metal sheet membranes of surface area packing density comparable to polymeric sheets is discussed to enhance ethanol dehydration energy efficiency and lower capital costs. The mechanical structural integrity of this new membrane product concept is substantiated with stress modeling. A relatively simple fabrication process that renders its low-cost manufacturing is delineated with experimental results. Separation process studies are conducted along with the membrane product development, which include (i) performance tests with real-world feeds and (ii) process design and simulation of membrane separation integrated with whole plants. The present membrane is highly permeable to water that significant one-pass dehydration can be obtained within a short residence time (seconds). However, the membrane performance can be drastically affected by the feedstock. Proper pre-treatment is necessary to achieve stable separation performances with the real-world feed. Based on experimental results, a two-stage membrane separation system is proposed to conduct ethanol enrichment and purification by replacing respective distillation and adsorption processes. More than 90% energy saving is shown using the membrane separation. Due to high permeance and high membrane area packing density, the membrane module estimated for a 40-million gal/year ethanol plant can be fairly compact, about the size of plate heat exchangers. This membrane technology is promising for retrofitting of existing corn ethanol plants and for building of new cellulosic ethanol plants in the future as well.

**(182f) Syngas Mass Transfer Analysis in a Hollow Fiber Reactor**

Orgill, J. J., Brigham Young University
Lewis, R. S., Brigham Young University
Atiyeh, H. K., Oklahoma State University
Originally presented on: 11/4/2013 17:05:00 - 17:27:00

Biofuel production via fermentation is produced primarily by fermentation of simple sugars. Besides the sugar fermentation route, there exists a promising alternative process that uses syngas (CO, $H_2$, $CO_2$) produced from biomass as building blocks for biofuels. Although syngas fermentation has many benefits, there are several challenges that still need to be addressed in order for syngas fermentation to become a viable process for producing biofuels on a large scale. One challenge is mass transfer limitations due to low solubilities of syngas species. The hollow fiber reactor (HFR) is one type of reactor that has the potential for achieving appropriate mass transfer rates for biofuels production, but a better understanding of mass transfer limitations associated with all syngas species is critical for assessing syngas usage during fermentation.

Mass transfer coefficients for $O_2$ and $CO_2$ have been typically assessed for HFRs using dissolved gas probes. In limited cases, CO has been studied using myoglobin assays. Though there are many reports on mass transfer coefficients in HFRs, studies are lacking that compare mass transfer coefficients of syngas species, including comparisons obtained from different measurement methods. Such comparisons are important to verify the accuracy of the experimental methods and to provide an increased understanding of the competing mass transfer limitations associated with syngas species.

In this study, we measured the mass transfer coefficient for $O_2$ in a PDMS hollow fiber reactor as a function of Reynolds number (flow rate) using a dissolved $O_2$ probe. The data was fit to a model characterizing $O_2$ mass transfer and the model was used to predict the mass transfer coefficients for CO and $H_2$. In order to verify the model predictions for these latter species, a second measurement technique using GC analysis was developed to measure CO and $H_2$ mass transfer coefficients. The GC technique was validated by comparing the mass transfer coefficient of $O_2$ from both GC and dissolved $O_2$ probe analysis. The mass transfer coefficients of both $H_2$ and CO in a HFR will be compared and the impact of such coefficients on syngas utilization will be presented. This study will greatly enhance the knowledge necessary to assess syngas usage and mass transfer limitations in a HFR used for syngas fermentation.

**(182g) An Innovative Methanol Synthesis Process Using Self-Heat Recuperation**

Kansha, Y., The University of Tokyo
Ishizuka, M., The University of Tokyo
Tsutsumi, A., The University of Tokyo

Originally presented on: 11/4/2013 17:27:00 - 17:49:00

The demand for methanol will continue increasing, since methanol is attracting as a fuel for fuel cells, and is an intermediate raw material of hydrogen and dimethyl ether (DME) which are categorized green energy source. As well as other chemical processes, many methanol synthesis processes have reactant recycle system with product separation due to the low conversion ratio of the reactor. For this product separation from reactant recycle stream, a gas-liquid separator or distillation process has often been used. However, this gas-liquid separator and distillation processes are well known as an energy consuming process. Thus, many chemical engineers and investigators managed to produce a catalyst which achieves high conversion from reactant to production in the reactor for whole process optimization.

Authors have developed self-heat recuperation technology, in which not only the latent heat but also the sensible heat of the process stream can be circulated without any heat addition, leading to reduction of energy requirement in several chemical processes. This self-heat recuperation technology is suitable for thermal and separation processes and is applied to several chemical processes for energy saving as case studies.

In this research, we investigated the feasibility of applying self-heat recuperation technology to the methanol synthesis process and developed an innovative process for methanol synthesis process from the energy saving point of view. By installing the self-heat recuperation technology to the methanol synthesis process, the energy consumption of the process can be greatly reduced. Thus, it can be said that installation of self-heat recuperation technology is the other option of process design method to achieve high conversion in the reactor.

**(253a) Molecular Simulations Of Ionic Liquids**

Zhang, S., Institute of Process Engineering, Chinese Academy of Sciences
Dong, K.,
Originally presented on: 11/5/2013 8:30:00 - 8:49:00

Molecular Simulations of Ionic Liquids

Suojiang Zhang ,Kun Dong

*Beijing Key Laboratory of Ionic Liquids Clean Process, State Key Laboratory of Multiphase Complex System, Institute of Process Engineering, Chinese Academy of Sciences, Beijing 100190, China.*

The complexity of interactions and diversity of combinations between cations and anions of have resulted in the further difficulty in applications of ionic liquids. The exploration of new functional ionic liquids is prompted to understand the structures and interactions in ionic liquids to correlate their properties.

In our group, a CUDA-based GPU program was developed to accelerate the computations of nano- and micrometer levels of ionic liquids. The volatile of ionic liquids, the interactions of ionic liquids and cellulose, and water, and $CO_2$, and the nano-aggragate in ionic liquids have been investigated by molecular simulations. We found that the ions are pairware in gas phase when aprotic ionic liquids were diluted into the gas. [1] The hydrogen bonds are important between anion and cellulose, but the role of cations can not been ignored in dissolving cellulose. [2] We also found the special stable structure of the hydrophobic [Bmim][$PF_6$] at [Bmim][$PF_6$]-water interface, and the water is difficult to hydrate the ions. In addition, it is found that the distribution of $CO_2$ molecules and $[Ppg]^+$ cations around [$BF_4$] anions are almost complementary when $CO_2$ molecules enter into [Ppg][$BF_4$] ionic liquids. In many neat ionic liquids and binary systems, ion-aggregate to form the nano-clusters is very common, in particular in the imidazolium-based ionic liquids with length of alkyl chain more than 4. [3]

[1] Kun Dong, Suojiang Zhang, et al. Phys. Chem. Chem. Phys., 2013, 15, 6034-6040.

[2] Yuling Zhao, Suojiang Zhang, et al. ChemPhysChem., 2012, 13, 3126– 3133

[3] Xiaomin Liu, Suojiang Zhang et al. J. Phys. Chem. B, 2012, 116, 4934-4942.

**(253b) Monitoring Inter- and Intra-Molecular Interactions in Imidazolium Protic Ionic Liquids HCnImNTf2 (n=0-12)**

Nikolakis, V., University of Delaware
Moschovi, A. M., FORTH/ ICEHT
Dracopoulos, V., FORTH/ ICEHT
Originally presented on: 11/5/2013 8:49:00 - 9:08:00

In this work we synthesized and studied the effect of the alkyl chain size in the structure of the imidazolium based Protic Ionic Liquids (PILs) 1-H-3-alkyl-imidazolium bis(trifluoromethanesulfonyl)imide $HC_nImNTf_2$ (n=0-12) , using vibrational spectroscopy (FTIR and Raman). The main characteristic of PILs is the presence of a proton on site 1 of the imidazolium ring. As a result PILs can form an extensive network of hydrogen bonds (HB).

The observed N-H vibrational modes in the high frequency spectral region at 3270-3150cm$^{-1}$ indicate that the liquid phase of ionic liquids consists of ion pairs. The observed redshifts, of the N-H and $C_{(i)}$-H bands frequencies (in the same region) with increasing alkyl chain length, suggest the strengthening of the HB between the anion and the cation. However, the $C_{(i)}$-H vibrational modes are primarily affected when n is varied between 0 and 3 (short alkyl chains), whereas N-H modes are affected for when n is varied from 3 to 12.Thus, the shift of the $C_{(i)}$-H mode can be attributed to the induction *+I* effect due to the presence of the alkyl chain in the ring, whereas the shift of the N-H modes is attributed to the intercationic interactions between the imidazolium rings that can be attributed to the parallel packing in the liquid phase structure.

The parallel stacking of the imidazole rings was also suggested from FT-Raman high frequency region. It was shown that when n>3, alkyl groups tend to develop structures similar to those of n-alkane melts, consisting of parallel imidazolium rings with oriented chains, defining polar and non polar regions. Moreover, the conformation of the anion $NTf_2^-$ was studied as a function of temperature and the alkyl chain length. It was shown that the elevation of the temperature favors the cis conformation of the anion, whereas trans conformation is dominant for bigger cations with long alkyl chains.

Finally, the enthalpy $\Delta H_{eq}$ of the conformational equilibrium of the anion $NTf_2^-$ for the PILs $HC_nImNTf_2$ (n=0-12) was experimentally calculated from spectroscopic data. The calculated enthalpy values reduce as the alkyl chain increases. The change in the slope of the calculated values from –ethyl to –propyl may reflect the change from intramolecular to intermolecular interactions.

**(253c) Beyond 1,3-Difunctionalized Imidazolium Cations: Properties of Ionic Liquids With Tri-, Tetra- and Pentafunctionalized Cations**

Bara, J. E., University of Alabama
Shannon, M. S., University of Alabama
Originally presented on: 11/5/2013 9:08:00 - 9:27:00

Although perhaps thousands of ILs have been reported in the literature, many are essentially built upon the same structure: 1,3-difunctionalized imidazolium cations. While these difunctionalized cations offer immense possibilities for tuning structure-property relationships, it is well-known that imidazolium-based materials can be further tailored via the introduction of various functionalities at the "2," "4," and/or "5" positions within the imidazole ring. However, strategies which employ trifunctionalized (or greater) imidazolium cations are seldom employed in materials design. Our group has synthesized a library of imidazole-based precursors via a scalable synthetic technique so as to further explore imidazolium cations with greater degrees of functionality, with focus on new solvents, polymers, composites, and liquid crystalline materials. This presentation will address the challenges and opportunities of "branching out" the imidazole ring so as to gain new dimensions for materials design and the potential implications in applications such as polymer membrane separations for energy applications (e.g. $CO_2$ capture) and self-assembly.

**(253d) Volumetric Properties of Mixtures of Ionic Liquids With CO2 Or CO2 + Ethanol**

Grandelli, H., Virginia Tech
Kiran, E., Virginia Tech
Originally presented on: 11/5/2013 9:27:00 - 9:46:00

We are exploring new approaches for potential use of ionic liquids in processing of natural polymers and specifically those that are important in processing of lignocellulosic materials. Our approach is to try to combine the desirable attributes of ionic liquids in terms of their solvent power with supercritical fluids to explore the extent to which dissolution levels can be tuned with pressure or fluid compositions. As an initial phase of this activity we have been evaluating the miscibility and volumetric properties (in particular the density and compressibility) of selected ionic liquids with $CO_2$ and $CO_2$ + ethanol. In this talk, we will present new data on the volumetric properties of two ionic liquids, namely [EMIM]Ac and [EMIM]Cl and their mixtures with carbon dioxide or carbon dioxide + ethanol, and comment also on the reactive interaction between [EMIM]Ac with $CO_2$. We will also report on the volumetric properties of mixtures containing Avicel as a model cellulose polymer at high pressures. The data are being generated in a special a high pressure variable-volume view-cell in which position of a movable piston is continually recorded in real-time while its position is altered at controlled rates with the aid of a motorized pressure generator. This position information is then converted to real time changes in the internal volume and hence the density as a function of pressure and/or temperature [1, 2]. The densities at a given temperature are then correlated with a polynomial function of the form $\rho = A + BP^{-1} + CP + DP^{-2} + EP^{m}$ were A, B, C, D, E and m are adjustable parameters. These functions were then used to determine the isothermal compressibilities for each mixture.

**Acknowledgment:** This research is supported by the National Science Foundation, CBET - Energy for Sustainability Program (CBET Award # 1250663).

[1] H. Grandelli and E. Kiran, Volumetric properties of mixtures of ionic liquid [EMIM]Ac with $CO_2$ as potential processing fluids for lignocellulosic materials at high pressures, Proceedings of the 3rd Iberoamerican Conference on Supercritical Fluids, *Prosciba 2013*, Cartagena, Colombia, April 1-5, 2013.

[2] J. M. Milanesio, J. C. Hassler, and E. Kiran, Volumetric properties of propane, n-octane an their binary mixtures at high pressures, Industrial and Engineering Chemistry Research, dx.doi.org/10.1021/ie4007084

**(253e) Molar Volume Dependence of N2 Solubility in Ionic Liquids**

Bennett, J.,
DeSilva, A., University of Notre Dame
Brennecke, J. F., University of Notre Dame
Originally presented on: 11/5/2013 9:46:00 - 10:05:00

Ionic liquids (ILs) have received increasing attention in recent years for their potential use in many different processes, particularly carbon capture. Many gas streams from which one might want to remove $CO_2$ also contain nitrogen. ILs that could be used for these separations include a diverse range of possible cation/anion combinations. It is therefore necessary to know the solubility of nitrogen in a broad range of ionic liquids. Unfortunately, there is a fair amount of disagreement amount the relatively limited $N_2$ solubility data available in the literature, most likely due to the challenges in measuring low solubility gases. Here we present nitrogen solubility data in a set of ionic liquids composed primarily of imidazolium cations with $BF_4$ and $Tf_2N$ anions, as well as several phosphonium and ammonium cations with $Tf_2N$ anions. The different ion combinations have molar volumes ranging from 150 $cm^3$/mol to 650 $cm^3$/mol. We have used a gravimetric microbalance produced by Rubotherm GmbH for solubility measurements at 60°C and pressures up to 150 bar. The resulting isotherms can be used to determine Henry's law constants for nitrogen in each IL. We will present results showing that the primary effect on the solubility of nitrogen in ILs is the molar volume of the liquid, with some differences between imidazolium and ammonium cations. Because of the low solubility of nitrogen in ILs and the time intensive nature of solubility experiments, the molar volume can be used as a quick estimate of the Henry's law constant for nitrogen in ionic liquids.

**(253f) Structure and Properties of Ionic Liquid/Organic Co-Solvent Mixtures**

Bandlamudi, S., South Dakota School of Mines & Technology
Cooney, M., University of Hawaii
Chupas, P.,
Benjamin, K. M., South Dakota School of Mines and Technology
Originally presented on: 11/5/2013 10:05:00 - 10:24:00

The extensive use of ionic liquids (ILs) is a significant achievement in extraction processes, especially in extracting lipids and proteins from biomass for bio-fuel production from algae and other bio-based feedstocks. The particular effectiveness of ionic liquids combined with polar-covalent molecules (PCM), such as methanol, as an extraction solvent mixture in extracting bio-oils from micro-algae has been demonstrated successfully by Young et al., 2010, the main highlight being auto-partitioning of extracted lipids into a separate phase. While the efficacy of these IL/PCM mixtures has been demonstrated experimentally, there is no current molecular-level understanding of the structural and thermodynamic properties of these mixtures.

The thermodynamic and structural properties of IL/PCM mixtures are explored using both molecular simulation and computational quantum chemical methods. Molecular dynamics (MD) simulations are conducted to predict mixture volumetric, thermal, and structural properties. MD simulations of thermodynamic mixing properties for IL/PCM binary mixtures show virtually ideal mixing behavior, while excess heats of mixing are predicted. In addition, MD predictions of structural properties like radial distribution functions and structure factors, supported by simulation visualization, provide information on local fluid structure and interactions between ionic liquid and polar-covalent molecules, including the change in ion aggregation with mixture composition. These simulation predictions are compared to small angle X-ray scattering (SAXS) and polarization data for model validation. Lastly, density functional theory (DFT) quantum chemical calculations are used to probe the shift in frequency of O-H stretch in methanol at various concentrations of IL/methanol mixtures, and compared to FTIR measurements.

These efforts provide an improved understanding between the molecular-level behavior and macroscopic thermodynamic properties of IL/PCM mixtures, to characterize these new solvents mixtures for separation applications.

**(253g) The Interactions of the Corn Protein Zein and Ionic Liquids**

Tomlinson, S., Kansas State University
Schlup, J. R., Kansas State University
Anthony, J. L., Kansas State University
Originally presented on: 11/5/2013 10:24:00 - 10:43:00

The solubility and secondary structure of the hydrophobic corn protein zein in ionic liquids was studied as a function of temperature and solvent properties. Â The solubility of zein was measured gravimetrically from 30Â°C to 60Â°C. Â The secondary structure of zein as a function of temperature and solvent was examined via curve fitting of infrared spectroscopic data. Â Two-dimensional correlation infrared spectroscopy was employed to monitor changes in the secondary structure of the zein as a function of temperature.Â These data clearly illustrate changes in the temperature dependent behavior between different solvents.

Most of the ionic liquids investigated change zein's secondary structure. Â Large solvent molecules increase the amount of β-turn secondary structure through non-polar interactions between the aromatic imidazolium cations (with alkyl substitutions) and the non-polar portions of the zein.Â Strong hydrogen bond accepting molecules increase the amount of β-turn secondary structure.

Solubility - solvent property relationships were explored to better understand the solvent properties of importance and how to design an ionic liquid to better dissolve a hydrophobic protein. It is found that a good solvent for zein has a small molecular volume, has a low polarity, and is a poor hydrogen bond acceptor. This combination of properties will enhance zein's solubility and limit secondary structure changes that can harm protein properties.

**(253h) Synthesis and Characterization of Novel Ternary Deep Eutectic Solvents**

Liu, Y., Donghua University
Xing, Y., Donghua University
Originally presented on: 11/5/2013 10:43:00 - 11:02:00

A series of novel, moisture-stable, ternary deep eutectic solvents (TDESs) has been prepared and characterized by mixing appropriate molar ratios of 1-butyl-3-methylimidazolium halide ([BMIM]X, X=Cl, Br), zinc halide ($ZnX_2$, X=Cl, Br) and various amides (acetamide, urea and N,N'-dimethyl urea). The synthesized ternary deep eutectic solvents are liquid at ambient temperatures and have interesting solvent properties. It shows that the most synthesized TDESs have much lower melting points than the conventional binary system according to the data of differential scanning calorimetry (DSC). Data on the melting point, viscosity and conductivity are presented and these are shown to be significantly different to other reported binary system. The [BMIM]Cl-$ZnCl_2$-$(Acetamide)_2$ system showed the lowest melting points (<-60°C), lowest viscosity and highest conductivity in all synthesized TDESs. The results showed that both viscosity and conductivity of all synthesized TDESs decrease with increasing temperatures and can be fitted by *Arrhenius equation* versus temperature. The data of conductivity and viscosity also fits well to the Walden's Rule. The influence of amides, halides and molar ratio of amides in ternary mixtures on the physical properties of the synthesized TDESs has been investigated.

**(311b) Systematic Molecular Simulation Investigation of the Influence of the Chemical Structure of Anion and Cation to the Properties of Imidazolium-Based Ionic Liquids and to the Thermodynamic Properties of Light Gases Dissolved Therein**

Economou, I. G., National Center for Scientific Research â€œDemokritosâ€
Androulaki, E., National Center for Scientific Research â€œDemokritosâ€
Vergadou, N.,
Originally presented on: 11/5/2013 12:50:00 - 13:10:00

This work focuses on the computational study of the physical properties of pure imidazolium-based tricyanomethanide Ionic Liquids (ILs), [$C_n$mim$^+$][$TCM^-$], and bis-trifluoromethylosulphonyl ILs, [$C_n$mim$^+$][$Tf_2N^-$], and on the prediction of solubility and diffusivity of light gases (e.g. Ar, $CO_2$, $N_2$) therein over a wide temperature range at atmospheric pressure. Molecular simulation plays a vital role in designing ILs with controlled permeability properties for use in a number of industrial processes, such as $CO_2$ capture, by revealing a wealth of microscopic information for the underlying mechanisms that are responsible for materials' macroscopic properties.

Here, long molecular dynamics (MD) simulations were performed in a wide temperature range and at atmospheric pressure in order to predict the thermodynamic, structural and dynamical properties of the pure ILs. A non-polarizable atomistic force-field was optimized in order to accurately predict the density and the self-diffusion coefficient. The dynamical heterogeneity exhibited by the ILs was investigated through the calculation of the non-Gaussian parameter and the deviation of the self-part of the van-Hove distribution function from the expected normal distribution. Subsets of "fast" and "slow" ions form individual clusters that consist of either mobile or immobile ions in the bulk. The structural relaxation process was analyzed by the self-part of the intermediate scattering function which exhibits strong wave number and temperature dependence. The distinct behavior of anions and cations at various length scales was thoroughly investigated as a function of the alkyl tail length. Detailed analysis of the ions' diffusion reveals preferential motion along the direction of the alkyl tail for the cation and the S-S vector for the [$Tf_2N^-$] anion. As the alkyl tail length increases, the heterogeneity in the dynamics becomes more pronounced and is preserved for several nanoseconds at low temperatures.

Simulation of light gases (e.g. Ar, $CO_2$, $N_2$) in the ILs at infinite dilution was performed in order to calculate the solubility and diffusivity of the gas molecules in the ILs. The solubility was estimated based on the Widom test particle insertion technique. In all cases, the agreement between experimental data and molecular simulation is excellent. The temperature effect was thoroughly studied for all the properties of the pure ILs and the dynamic properties of light gases dissolved in the above mentioned systems. Moreover, the influence of the cation's alkyl chain length and the choice of the anion were investigated in order to study the underlying molecular mechanism that governs the macroscopic behavior of these systems. Diffusivity was estimated through the mean square displacement of the ions.

In conclusion, strong spatial organization with an indication of alkyl tail aggregration phenomena was determined. This organization leads to an inhomogeneous bulk structure and is connected to heterogeneity in the dynamics of these systems which has been also reported by previous studies of similar systems. This inhomogeneity affects the macroscopic physical properties of the system.

**(311c) Distinctive Nanoscale Organization of Double- Cationic Versus Single-Cationic Ionic Liquids**

Li, S., Vanderbilt University
Feng, G., Vanderbilt University
Banuelos, J. L., Oak Ridge National Lab
Rother, G., Oak Ridge National Lab
Fulvio, P. F., Oak Ridge National Lab
Dai, S., Oak Ridge National Laboratory
Cummings, P. T., Oak Ridge National Laboratory
Originally presented on: 11/5/2013 13:10:00 - 13:30:00

Double-cationic ionic liquids are attracting the increasing research interests recently. In this work, the distinctive structural organization of double-cationic ionic liquids (DILs) with varying alkyl linkage chain lengths is systematically investigated using classical molecular dynamics (MD) simulations. In comparison with their single-cationic ionic liquids (SILs) counterparts, DILs with varying length alkyl chains exhibit structural features that are different from SILs regardless of anion types. SILs are found to exhibit various structural heterogeneities that are more pronounced than DILs. Moreover, the predominant role of anion type in the structure of DILs was verified be analogous to that in MILs. In the end, the different nanoscale organizations of DILs and MILs are rationalized by straight- and folded-chain models proposed for the nanoaggregates in DILs and those in MILs. This study provides molecular insights into the nature of structural heterogeneity in DILs as well as a theoretical model for the interpretation of experimental observations.

**(311d) Aggregation in Ionic Liquids: Molecular Dynamics Simulation**

Liu, X., Institute of Process Engineering, Chinese Academy of Sciences
Zhang, X., Institute of Process Engineering, Chinese Academy of Sciences
Zhang, S., Institute of Process Engineering, Chinese Academy of Sciences

Originally presented on: 11/5/2013 13:30:00 - 13:50:00

The spatial microheterogeneity in ionic liquids (ILs) is one of the the key features and has been proved useful in understanding the special behavior and interpreting many Physical phenomena of ILs, including the heterogeneous self diffusion, surface layering, as well as their surfactant-like micelles formed in IL/water mixtures.[1,2] As a result, much attention has been paid to the related researches, including both experiment and simulation. [3,4]

The self-aggregation or segregation exists both in IL solution and neat IL composed entirely of ions. In this work, the self-aggregation in ILs composed by $[C_n mim]^+$ (n=2, 4, 6, 8 , and 10) and $Br^-$ were investigated by molecular dynamics simulation at the molecular level based on the all-atom force field. Both the neat ILs and the mixture of $[C_{12}mim]Br$ with water were studied. In order to reduce the computational time, united atom force field was utilized for the mixture of $[C_{12}mim]Br$ with water. Center of mass and site-to-site radial distribution functions (RDFs) were employed to study the ionic pair interactions. To depicture more visual structures of these ILs, the spatial distribution functions (SDFs) that stand for the three-dimensional probability distributions of ions and molecule were studied. A way based on the atomicity density was developed for estimating and understanding the inhomogeneous structure at nanometer scale. Dynamics behaviour and diffusion of ions, molecules and clusters were characterize through the mean-square displacement and self diffusion fuctions.

**References:**

(1) Wang, Y.; Jiang, W.; Yan, T.; Voth, G. A. *Acc. Chem. Res.* **2007**, *40*, 1193-1199.

(2) Dupont, J. *Acc. Chem. Res.* **2011**, *44*, 1223-1231.

(3) Chen, S. M.; Kobayashi, K.; Kitaura, R.; Miyata, Y.; Shinohara., H. *ACS Nano* **2011**, *5*, 4902-4908.

(4) Liu, X.; Zhou, G.; Zhang, S.; Wu, G. *Mol. Sim.* **2010**, *36*, 1123-1130.

**(311e) Modeling Chemical and Physical Interactions in CO2/Ionic Liquids With the Reaxff Force Field**

Zhang, B., US DOE-National Energy Technology Laboratory
van Duin, A. C. T., Pennsylvania State University
Johnson, J. K., University of Pittsburgh
Originally presented on: 11/5/2013 13:50:00 - 14:10:00

Ionic liquids (ILs) are promising sorbents for $CO_2$ due to their essentially zero vapor pressure and the potential to tune the $CO_2$-IL interaction. Both experimental and computational studies of $CO_2$/IL systems have been carried out. The ILs can interact with $CO_2$ through one or more of the following: physical absorption, complexation, chemical reaction. It is a challenge for a computational model to be able to accurately account for all three of these types of interactions. Physical interactions can be described with classical potentials very well, but cannot account for the bond formation and breaking events associated with chemical interactions between $CO_2$ and ILs. Treatment of complexation and chemical reactions usually requires high-level quantum chemical methods that are computationally expensive. An alternative approach is to use a reactive force field that can capture both physical interactions and chemical reactions. We have developed a new parameterization of the ReaxFF reactive force field for tetrabutylphosphonium glycinate ($[P(C_4)_4][Gly]$) and $CO_2$. We developed an extensive training set for both the pure $[P(C_4)_4][Gly]$ system and the $[P(C_4)_4][Gly]/CO_2$ mixture from periodic and gas phase density functional theory (DFT) calculations. We have validated the optimized ReaxFF force field by comparing the energies from DFT and ReaxFF calculations for the structures in the training sets. We also tested the predictability of the optimized force field by comparing the same configurations of ILs with $CO_2$ not in the training sets. The trends of relative stability of different configurations based on energies from DFT and ReaxFF methods are in good agreement. The optimized ReaxFF force field has been used in large-scale molecular dynamics (MD) simulation. The calculated density is in very good agreement with experiment measured values at 300 K at ambient pressure. The radial distribution function and transport properties (MSD, diffusivity) of pure IL and IL with $CO_2$ mixture have been calculated and compared with those from classical potential. The reaction between $CO_2$ and IL has also been monitored and compared with first principles MD results.

**(311f) Prediction of Thermodynamic Properties of Ionic Liquids: Improved Performance From Ab Initio Molecular Dynamics**

Zhang, Y., University of Notre Dame
Maginn, E. J., University of Notre Dame
Originally presented on: 11/5/2013 14:10:00 - 14:30:00

Classical molecular dynamics and Monte Carlo simulations have been used extensively over the past ten years to predict the thermodynamic and transport properties of ionic liquids. As more experimental data for these systems becomes available, it is becoming clear that the quality of the force field is essential for making accurate property predictions. In this talk, we show how relatively short ab initio molecular dynamics (AIMD) simulations can be used to improve the accuracies in the prediction of the melting points and electrochemical windows of ionic liquids. On one hand, AIMD was used to better parameterize classical force fields, which can then be used to conduct traditional classical simulations on large systems for long time scales. We show that properties such as melting point can vary tremendously depending on both the magnitude and distribution of partial charges. On the other hand, AIMD was directly combined with classical MD simulations to provide better description of the potential energy surface. It was found that a quick relaxation of snapshots from classical MD simulation using AIMD can improve the calculated electrochemical window significantly.

**(311g) Molecular Thermodynamic Characterization of Ionic Liquids for Separation Processes**

Vega, L. F., MATGAS Research Center and Carburos MetÃ¡licos, Air Products Group
Llovell, F., MATGAS Research Center
Oliveira, M. B.,
Vilaseca, O., MATGAS Research Center and ICMAB-CSIC
Marcos, R. M., Universitat Rovira i Virgili (URV)
Coutinho, J. A. P., Universidade de Aveiro
Originally presented on: 11/5/2013 14:30:00 - 14:50:00

Ionic Liquids (ILs) have been widely studied in the last ten years because of their potential applications from a technological point of view as well as their highly non-ideal properties from a fundamental perspective. The fact that they are liquid and have nearly zero vapor pressure values at room temperature, as well as their capacity to modify their adsorption capabilities by modifying the anion-cation pair, convert them in good candidates for gas absorption separation and several other industrial processes. However, the process of thermodynamic characterization, needed before put into a final application, is very expensive and time-consuming in terms of experimental work. Hence, different equations of state have been applied to the estimation of the thermodynamic properties of these ionic liquids, with different degrees of success. The complexity of these molecules has reduced the ability to reproduce some thermophysical properties in good agreement with the measured data, and most of the work done up to now with success has focused only in the description of the temperature-density diagram and the solubility of a particular gas on the ionic liquid.

One of the challenges in these theoretical approaches is to make them as general as possible, reproducing several features of ionic liquids and, in particular, their interfacial and derivative properties, as well as mixture data, with the same degree of accuracy as temperature-density data, instead of focusing on a particular property.

We will present and discuss the work our team has been doing in this field in the last years. This work is intended to fill this gap by proposing and validating a general and simple molecular model with the soft-SAFT equation of state [1] giving physical insight into the development of the model and the accuracy of the parameters for predictive purposes. This is accomplished by including different calculations of thermophysical properties of ionic liquid families with two different cations (imidazolium and pyridinium) and several anions ($BF_4$, $PF_6$, $Tf_2N$, $MeSO_4$, etc.). In all cases, simple but reliable molecular models are presented based on structural information and, also, in results obtained from molecular simulations and ab-initio calculations. As a first step the molecular parameters are fitted to single-phase equilibrium data and used in a transferable manner to predict other properties, such as high pressure density data, enthalpy of vaporization, critical points, interfacial tensions [2] and second order thermodynamic derivative properties. Correlations with the molecular weight are identified. Some derivative properties (heat capacity, isothermal compressibility, etc.) are used to discriminate among sets of parameters. The second step involves the assessment of the predictive ability to reproduce the solubility of several pollutant gases ($CO_2$, $SO_2$, $NH_3$, $H_2S$) in those ILs, as compared to experimental data [3-5]. In most of the cases, excellent agreement is found using a single temperature-independent binary parameter. The transferability of the binary parameters is also assessed. An estimation of the energetic properties is given in order to determine the effect of enthalpic and entropic effects on the solution. Finally, aqueous mixtures of ionic liquids and their mutual solubilities (liquid-liquid equilibrium) are reproduced with this approach [5-6].

***Acknowledgements***

*F. Llovell acknowledges a TALENT fellowship from the Generalitat de Catalunya. M.B. Oliveira acknowledges her Post-Doctoral grant (SFRH/BPD/71200/2010). This work has been partially financed by the Spanish government (projects CTQ2008-05370/PPQ and CENIT SOST-CO2, CEN2008-01027), the Catalan government (project 2009SGR-666), the Portuguese Government (project PTDC/EQU-FTT/102166/2008) and from Carburos Metálicos, Air Products Group. CICECO is being founded by FCT through project Pest-C/CTM/LA0011/2011.*

**References**

[1] F.J. Blas, L.F Vega, Mol. Phys. 92 (1997) 135.

[2] J.D. van der Waals, Z. Phys. Chem. 13 (1894) 657.

[3] J.S. Andreu, L.F. Vega, J. Phys. Chem. B 112 (2008) 15398.

[4] F. Llovell, R. M. Marcos, N. Mac Dowell, L. F. Vega, J. Phys. Chem. B. 116 (2012), 7709.

[5] M. B. Oliveira, F. Llovell, J.A.P. Coutinho, L. F. Vega, J. Phys. Chem. B. 116 (2012) 9089.

[6] F. Llovell, E. Valente, O. Vilaseca, L.F. Vega, J. Phys. Chem. B 115 (2011) 4387.

**(367a) Phase Behavior of Gases in Ionic Liquids**

Shiflett, M. B., DuPont Company
Yokozeki, A., DuPont
Originally presented on: 11/5/2013 15:15:00 - 15:35:00

This presentation will emphasize the importance of characterizing the global phase behavior of gases in ionic liquids and how this can provide insight into new applications. Solubility measurements of several gases in ionic liquids are discussed and important experimental details are highlighted when measuring the vapor-liquid equilibria using a gravimetric microbalance and vapor-liquid-liquid equilibria using a mass-volume technique. The global phase behavior has been well correlated with a modified cubic equation and many of the systems described exhibit type-III and type-V phase behavior according to the van Konynenburg-Scott classification. We have found that gases can exhibit different solubility behaviors in ionic liquids (*i.e.* physical and chemical absorption) and that these behaviors can be analyzed with the present equation of state using a simple association model and excess thermodynamic functions. Gas separation using ionic liquids is discussed which includes examples of ternary phase behavior (experiments and model calculations) and what will be required to develop future commercial processes.

**(367b) The Viscosity of Ether-Functionalized Ionic Liquids**

Chen, Z., Nanyang Technological University
Xue, T., Nanyang Technological University
Lee, J. M., Nanyang Technological University

Originally presented on: 11/5/2013 15:35:00 - 15:55:00

The development of low viscous and highly conductive ionic liquids is always being driven by the ambition to improve their performance in related fields, like heterogeneous catalysis and supporting electrolyte. Generally, a low-viscosity ionic liquid is characteristic of well-delocalized positive and negative charges, low-symmetry ions and highly flexible side chains, for example ether groups. Recently, hole theory indicated that: 1) the ions in ionic liquids cannot pack closely and lots of holes may come into existence; 2) the bulk volume consists of the inherent volume and the total volume of holes; and 3) a large amount of total hole volume is of advantage to a high rate of mass transfer and low viscosity.[1,2]

In this study, 12 ether-functionalized ionic liquids and their alkyl counterparts short of an alkoxy O atom were prepared and characterized to determine the reason for the alkoxy chain effect on viscosity, as shown in Figure 1, based on the ether chains' contributions to viscosity ($\eta_O/\eta$), the activation energy of viscosity ($\Delta E_\eta$) and the total free volume ($\Delta FV_O$), which was calculated from density and surface tension [$FV = 2N \times 0.6791(kT/\gamma)^{3/2}$].[2] For example at 25 °C, when the $C_3H_7$- group in [$C_3H_7$-MPyrr]$NTf_2$ was replaced by $C_2H_5OCH_2$-, the viscosity and its activation energy would be reduced from 62.4 cP and 28.83 kJ $mol^{-1}$ to 38.7 cP and 25.29 kJ $mol^{-1}$ respectively, yielding $\eta_O/\eta = 0.620$ (38.7/62.4) and $\Delta E_\eta = -3.54$ kJ $mol^{-1}$ (25.29 - 28.83) respectively, accompanied by $\Delta FV_O = 4.76$ mL $mol^{-1}$, while the total free volume in [$C_2H_5OCH_2$-MPyrr]$NTf_2$ is 32.25 ml $mol^{-1}$. Although the obtained $\eta_O/\eta$, $\Delta E_\eta$ and $\Delta FV_O$ for the 12 pairs of ionic liquids are highly variable, both $\eta_O/\eta$ and $\Delta E_\eta$ show a decreasing dependence on $\Delta FV_O$. As a result, the alkoxy chain effect on reducing viscosity was ascribed to the ability of the highly flexible alkoxy chains to increase the total free volume for the convenience of mass transfer. In order to maximize the effect on decreasing viscosity, the alkoxy group was supposed to contain a rod-like alkyl tail and a spacer length as short as only one methylene between O atom and the cationic N atom, and be in a non-hydrogen-bonding environment.

[1] A. P. Abbott, *ChemPhysChem*, **2004**, *5*, 1242-1246.

[2] J. Z. Yang, X. M. Lu, J. S. Gui and W. G. Xu, *Green Chem.*, **2004**, *6*, 541-543

[3] Z. J. Chen, T. Xue, J. M. Lee, *Rsc Adv.*, **2012**, *2*, 10564-10574.

**(367d) Time-Dependent Diffusion and Chemical Exchange in Mixtures of An Amine-Functionalized Ionic Liquid and Carbon Dioxide By NMR**

Hazelbaker, E., University of Florida
Wang, H., University of Florida
Budhathoki, S., University of Notre Dame
Shah, J. K., University of Notre Dame
Maginn, E. J., University of Notre Dame
Vasenkov, S., University of Florida
Originally presented on: 11/5/2013 16:15:00 - 16:35:00

Task-specific ionic liquids (TSILs) containing amine functional groups are promising materials for carbon dioxide capture. In this work a novel diffusion NMR technique, which combines advantages of high magnetic field and high magnetic field gradients, was applied to investigate self-diffusion in mixtures of an amine-functionalized TSIL and carbon dioxide. The measured diffusion data reveal time-dependent diffusivities of carbon dioxide in the mixtures on the millisecond time scale. This dependence was attributed to the existence of chemical exchange of carbon dioxide, i.e. exchange of carbon dioxide between the chemisorbed and physisorbed states in the mixtures under conditions of diffusion studies. At the same time, ion diffusivities were found to be diffusion time independent. High field exchange NMR was used to confirm the existence of the chemical exchange process and to obtain the mean residence times of carbon dioxide in the chemisorbed and physisorbed states. The diffusion and exchange NMR studies were performed for different molar ratios of carbon dioxide and at different temperatures. The experimental results will be compared with the predictions of computer simulations performed for the same or similar mixtures of TSIL and carbon dioxide.

**(367e) Investigation of the Feasibility of Flue Gas Separations Using An Ionic Liquid From a Thermodynamic Point of View**

Raeissi, S., Shiraz University
Florusse, L. J., Delft University of Technology, Faculty of Applied Sciences, DelftChemTech (PCMT)
Peters, C. J., Petroleum Institute
Originally presented on: 11/5/2013 16:35:00 - 16:55:00

Just one decade ago, the term "ionic liquid" was unknown to the majority of chemical engineers. Today, this unique family of substances is being investigated so intensely that over four-thousand journal articles are being published on ionic liquids in a period of just one year. One of the areas that have attracted great research interest is the system of $CO_2$ and an ionic liquid (IL). A Scopus search on the combination of words "ionic liquid" and "carbon dioxide" leads to 250 results for the year 2012. Potential applications involving such systems include $CO_2$ absorptions and separations by ionic liquids. Due to their importance in the design of such processes, numerous publications have appeared, and are continuing to do so, on the phase behavior of $CO_2$ + ionic liquid systems. However, the number of articles that present the solubility of carbon monoxide (CO) in ILs is an order of magnitude less than the corresponding papers on $CO_2$. While the solubility of CO in ILs is not as high as that of $CO_2$, it is still significant enough to deserve further investigation. Potential applications include, for example, IL-enhanced reactions involving CO, or gas purifications with ILs. Flue gas, in particular, consists mainly of $N_2$, CO, $H_2$, $CO_2$, $SO_2$, and $N_2O$. With environmental laws becoming increasingly strict, the purification of gas effluents is gaining even more urgency. In this respect, various routes have been investigated. One potential path that can be investigated regarding its feasibility, is the separation of flue gas pollutants such as CO, $CO_2$, $SO_2$, $N_2O$, and $CH_4$ by ionic liquids, for example, in the form of supported ionic liquid membranes.

In this respect, this study investigates the high-pressure solubility of CO in 1-butyl-3-methylimidazolium bis(trifuoromethylsulfonyl)amide ([bmim][$Tf_2N$]) up to temperatures and pressures of 460 K and 10 MPa, respectively, using a synthetic method. The results obtained show that the extent of solubility of CO in this ionic liquid is not great. For example, the single gas solubility of CO at 323 K and 2 MPa is about 0.02 in mole fraction, while the corresponding values for $CO_2$, $N_2O$, and $CH_4$ at the same temperature and pressure are 0.3, 0.3, and 0.04, respectively. A very rough estimate indicates that at 323 K and 2 MPa, the ratios of single-gas equilibrium molar concentrations in [bmim][$Tf_2N$] of carbon dioxide to nitrogen is roughly $CO_2/N_2 \approx 17$, and the corresponding values for other gases are: $SO_2/N_2 >> 17$, $N_2O/N_2 \approx 17$, $CO/N_2 \approx 1$, $CH_4/N_2 \approx 2$. The solubilities of the different gases are higher at higher pressures, much more so for the highly soluble gases such as $SO_2$, $N_2O$, and $CO_2$ and much less for the low-soluble gases such as $N_2$ and $O_2$.

Therefore, flue gas purifications can be more efficient at higher pressures. Since the above estimates indicate single-gas solubility differences of up to one order of magnitude with one another, it should be possible, at least from a thermodynamic point of view, to absorb some of the impurities of flue gases in this ionic liquid in order to separate them from $N_2$ and $O_2$, before flue gas is vented into the atmosphere.

However, if one focuses on CO capture alone, the results of this study may seem discouraging at a first glance. Although CO was shown to dissolve better in [bmim][$Tf_2N$] than in most conventional organic solvents, it still does not have the high solubility that is desired in order to achieve easy and cheap separations. However, if one focuses on the common liquid absorbents that have been investigated for CO separations, it is seen that they are not capable to simultaneously remove other impurities from flue gas. On the other hand, ionic liquids are very good absorbents for $CO_2$, $SO_2$, and $NO_x$. Organic solvents are simply not good solvents for volatile gases such as CO, therefore to increase the solubility in a liquid, researchers have mainly focused on absorbents which form complexes with CO. Examples include COSORB, ammoniacal copper liquor, Fe(TIM), palladium complex, and selenium and secondary amine. However, such complex-forming absorbents do not make complexes with other flue gas pollutants. For example, in the COSORB process $CO_2$ and $CH_4$ are chemically inert to the solvent, $CO_2$ does not react with the Fe(TIM) solvent, the palladium complex is good for separating $CO_2$ and $CH_4$ from CO, and the selenium and secondary amine system can be used to separate $CH_4$ from CO. Therefore, using such absorbents for CO removal would require additional separation units in order to remove the other impurities of flue gases. But ILs can simultaneously dissolve many gases, and additionally, since gas solubility in an IL is a purely physical process with no chemical interactions involved, the ILs can be cheaply and very easily regenerated by pressure drop or temperature increase. One further advantage of ILs, compared to other liquid absorbents, is that upon regeneration, the separated gases are not polluted with the solvent, because an ionic liquid has insignificant vapour pressure. Because of all these considerations, flue gas purifications with ionic liquids deserve further investigations. However, it is suggested that future investigations should consider mixed gases and real flue gases instead, since the presence of other solutes in an IL have been shown to increase/decrease the extent of solubilities compared to their corresponding binary-pair values, i.e., to show cosolvency or antisolvency effects.

**(367f) Free Volume Basis of Gas Solubility and Selectivity in Functionalized Imidazolium-Based Ionic Liquids**

Shannon, M. S., University of Alabama
Bara, J. E., University of Alabama
Horne, W. J., University of Alabama
Originally presented on: 11/5/2013 16:55:00 - 17:15:00

Due to the tunable structures and functionalities of ionic liquids (ILs), current research efforts have introduced several unique materials for industrial applications, including solvents and media for $CO_2$ capture. Fundamental conclusions and aspects of these IL materials and their physical properties are imperative to understanding governing driving-forces for gas separations and underlying molecular structures. Recently, we investigated fractional free volume (FFV)-dependence of gas solubility in 165 existing and theoretical 1-*n*-alkyl-3-methylimidazolium-based ionic liquids (ILs) via COSMOtherm, a very powerful and rapid software utilized commonly now for predicting IL thermophyscial properties.

Our present study investigates the effects on gas solubility of novel functional groups by replacing *n*-alkyl chains with nitriles ($C_nH_{2n}C{\equiv}N$) and ethers or oligo(ethylene glycols) $(-C_2H_4O-)_n$. Initial computational results with these new functional groups via COSMOtherm reflect deviations in FFV from previously studied [$C_n$mim] ILs. Polar substituents provide a higher $CO_2$ binding affinity and repulsion of non-polar gases (i.e. $CH_4$, $N_2$, etc.) leading toward improved $CO_2$ separation performance. Initial results suggest decreased FFV values for appended nitrile and ether groups when compared to previous [$C_n$mim] IL data, which would also imply decreased gas solubility. However, with this observed decrease in FFV due to chain aggregation of polar entities, ideal gas selectivity is shown to increase with these materials being more size selective within these microdomain/voids and unable to accommodate the larger $CH_4$ and $N_2$ molecules.

Previous models presented for gas solubility in ILs consist of a molar volume-basis for families of 1-*n*-alkyl-3-methylimidazolium ILs. Free volume within these ILs as well as the currently studied nitrile- and ether-based ILs was shown to be the underlying property driving gas solubility and selectivity. For light gases (i.e. $CH_4$ and $N_2$) considered in $CO_2$ separations, solubility increases with increasing FFV implying direct proportionality with free volume. However, solubility of $CO_2$ was observed to decrease with increasing FFV, with $CO_2$ selectivity being a function of free volume to the (-3/2) power. Controlling IL FFV via cation/anion combinations, as well as, cation functional groups can lead to improvements in $CO_2$ solubility and selectivity and permeability in IL-based membranes.

**(367g) The Effects of Water Concentration On the Properties of Imidazolium-Based Ionic Liquids**

Niazi, A. A., RWTH Aachen University
Crothers, A., RWTH Aachen University
Rabideau, B. D., RWTH Aachen University
Ismail, A. E., RWTH Aachen University
Originally presented on: 11/5/2013 17:15:00 - 17:35:00

In the conversion process of production of tailor-made biofuels from waste biomass, cellulose is solvated and broken down into glucose units using solvents such as ionic liquids (ILs). Because this process releases trapped water molecules from the cellulose matrix, and because water is known to inhibit the dissolution of cellulose in IL's, study of the effect of water content on local interactions between individual components of the system has an important role in understanding the details of the mechanism of cellulose dissolution using ILs. In this work, we have analyzed the results of series of molecular dynamics (MD) simulations of twelve different commercially available imidazolium-based ILs: a combinatorial set of four cations: [dmim], [emim], [pmim], and [bmim] and three anions: [DMP], Cl, and [Ac]. For each IL, we have tested the effect of presence of water molecules on structural properties and dynamics of systems with water mole fractions ranging from 0.0% to 99.9%. Particularly, we have examined bulk properties (density), structural properties (radial distribution functions, water clustering and hydrogen bonding) and dynamical properties (shear viscosity and diffusion coefficients) of those mixtures and have made direct comparisons with experimental works whenever possible. We have observed for each IL a critical concentration, usually at water mole fractions between 70% and 80%, at which the mixture undergoes a transition from water dispersed in IL to an aqueous solution.

**(387a) Findings On Life Cycle Aquatic Ecotoxicity Impacts of Ionic Liquids**

Mehrkesh, A., University of Colorado Denver
Karunanithi, A. T., University of Colorado Denver
Originally presented on: 11/5/2013 15:15:00 - 15:34:00

Ionic liquids (ILs) are a class of chemicals that are being developed as a replacement for volatile organic solvents for a range of industrial applications (e.g. $CO_2$ capture, aliphatic/aromatic separations, battery electrolytes, cellulose dissolution). Although ILs can potentially decrease health impacts associated with air emissions due to their nonvolatile nature, many of them are highly soluble in water and can become persistent aquatic pollutants. Environmental impacts of tailored ionic liquids arise both from their direct release into the environment during or after use phase as well as indirect emissions during their synthesis. In this work, we use the USEtox model to quantify aquatic ecotoxicity impacts over the life cycle of an Ionic liquid (IL), 1-butyl-3-methylimidazolium bromide ($BMIM^+ Br^-$), from cradle to grave. The aquatic ecotoxicity impacts from synthesis of this IL as well as its release into the enviorment have been taken into account. Life cycle impact assessment methods define environmental impacts in terms of characterization factors. In this work, for the first time we have developed these characterization factors for ILs. The USEtox, a multimedia fate and transport model, was used to derive fate factors and exposure factors needed for this calculation while a large number of literature data related to toxicity of ILs in a variety of species was used to derive the effect factor. Finally we compare the ecotoxicity production impacts with that of direct exposure.

**(387b) Comparative Life Cycle Assessment of Vaccine Cold Chain Infrastructure in Developing Countries**

Mousavi, B., University of Colorado Denver
Karunanithi, A. T., University of Colorado Denver
Originally presented on: 11/5/2013 15:34:00 - 15:53:00

An important component of vaccine programs in low-and middle- income countries is the cold chain, which is a collection of cold rooms, refrigerators, freezers and cold boxes that keep vaccines at an appropriate temperature. Vaccines require cold storage (between $2^o$ C and $8^o$ C) in order to remain stable and be potent when administered. Heat sensitive vaccines lose potency after exposure to temperatures above the recommended range and freeze sensitive vaccines lose potency when exposed to freezing temperatures. In the next few years there is an urgent need to streamline and expand efficient and reliable cold chain infrastructure in these countries. In this context, the importance of designing and building new cold chain infrastructure that minimizes the environmental impact of energy, materials and processes both within the country and globally cannot be overstated.

In this study a complete life cycle assessment (LCA) analysis for 5 different types of vaccine refrigerators is done for the case of ghana. All of these refrigerators work in regions where accessibility to grid electricity is very minimal. Firstly, the cradle to grave life cycle inventory for each case and secondly the complete analysis of Global Warming Potential (GWP), Human Health (HH), acidification, ozone depletion and smog impacts are presented. After this LCA analysis, policy makers in developing countries can decide to choose the best option (eco-friendly) based on any environmental impacts.

The 5 different types of vaccine refrigerators selected for this study are:

- solar powered battery supplemented refrigerator (SPBS)
- Solar Powered battery-free refrigerator (SPBF)
- Fully Kerosene refrigerator (FK)
- Hybrid Kerosene- electricity refrigerator (HKE) (biomass cogeneration generates electricity)
- Fully LPG Refrigerator (FL)

The SPBS and SPBF systems work in Vapor-Compression cycle refrigeration system and the other types work in absorption refrigeration cycle. All of these types are specific products from famous companies that produce vaccine refrigerators for developing countries all around the world.

The functional unit of this study was set as providing 1 liter cold storage for vaccines during one year. The life span time for SPBS and SPBF was 15 years and for FK,HK and FL is 5 years.

The cradle to grave unit processes of this study include: refrigerant production, insulation foam production, other materials production, photovoltaic modules production, battery production, assembly of all parts of a refrigerator, transportation, LCI of kerosene and LPG from extraction to combustion and discharge of refrigerant and insulation foam to air at the disposal phase.

The Impact assessment results show that FK and SPBF vaccine refrigerators have respectively the highest and lowest GWP and HH impacts. On the other hand, if the HKE used instead of FK all of the environmental impacts decrease dramatically.

**(387c) Hybrid Life Cycle Assessment Model of Silicon Photovoltaics**

Yao, Y., Northwestern University
Chang, Y., Northwestern University
Masanet, E., Northwestern University
Originally presented on: 11/5/2013 15:53:00 - 16:12:00

Research on sustainable energy resources is receiving increasing attention because of growing concern about severe pollution associated with fossil-fuel power plants. Solar energy is one of the most promising alternatives. Photovoltaic (PV) technology is gradually becoming competitive in power generation market for its abundant resources and affordable electricity[1]. Among all solar technology, silicon PV technology has been successfully commercialized and has the greatest market share[2]. However, any energy production will also consume energy, generate pollutants and result in some environmental impacts [3]. In order to assess the energy sustainability and environmental impacts of silicon PVs, a lot of work has been done on applying Life Cycle Assessment (LCA) tools to solar energy. Currently, the majority of assessments adopt process-based LCA[1] which is a bottom-up method that is limited by system boundary variations, difficulty for replicating results, and inability to provide assessments for future products due to frequent use of proprietary data. Constrained by incomplete system boundary, results of process-based LCA studies often fail to evaluate the entirety of energy and environmental impacts of solar technology on the whole society and ecosystem. Another tool for LCA studies is Economic Input-output LCA (EIO-LCA) model which has a more comprehensive system boundary, allows for comparisons in systems-level and capacity of assessing future product developments[4]. However, EIO-LCA has time lag for current practices, highly depends on government statistical data, and suffers from sectorial aggregations[5].

In this paper, a hybrid LCA model for silicon PVs, which takes advantage of both process-based method and EIO-LCA, was developed. This hybrid model expanded the system boundary to all supply chain materials of silicon PV by matrix operation (e.g. usage of petroleum coke to produce silicon carbide for wire sawing, electricity generation and usage of all upstream materials). Another strength of this hybrid LCA model is that the geographical shift of PV manufacturing, such as expanding Chinese silicon PV production, can be easily estimated by using publicly accessible input-output data of that country. Since silicon PV has been commercialized for several decades, the variation of technology parameters caused by geographic shift is not significant[1]. However, the difference of sustainability performance of PV made in different countries can still be considerable because of diverse economic structure and energy mix. This difference can be captured and assessed by this hybrid LCA model.

For model demonstration, a case study of U.S. multicrystalline PV production was conducted. Results show that the GHG emissions and energy intensity are 41 g $CO_2$-eq/kWh (based on 1,800 kWh/$m^2$/year, 14% module efficiency, 0.75 performance ratio and 30 years lifetime) and 5811 MJ/$m^2$ (balance of system is included), respectively. The Energy Payback Time (EPBT) is calculated to be 2.54 year. Results of this study are 30 – 60% higher than previous studies using process-based LCA model after harmonization[1]. This is caused by the comprehensive boundary of hybrid model, covering more upstream materials in the supply chain and thus more GHG emissions and energy considerations in research system. Sensitivity analysis was conducted to filter dominant factors influencing the sustainability of multicrystalline PV made in the U.S. from a more comprehensive view, which can provide useful recommendations for future research. Given that this model also shows specific energy consumption and GHG emissions of all sectors in U.S. economy caused by multicrystalline PV production and operation, study findings provide policy maker and general public with a more holistic view of the impacts of PV technology over the whole society and ecosystem. More case studies for silicon PV made in other countries, such as China, will be shown in this paper to give a comparison and show the fluctuation of the sustainability performance of silicon PV caused by geographic shift.

**References**


[1] D. D. Hsu, P. O'Donoughue, V. Fthenakis, G. A. Heath, H. C. Kim, P. Sawyer, *et al.*, "Life Cycle Greenhouse Gas Emissions of Crystalline Silicon Photovoltaic Electricity Generation," *Journal of Industrial Ecology,* vol. 16, pp. S122-S135, 2012.

[2] "Solar Generation 6," European Photovoltaic Industry Association (EPIA)2011.

[3] EPA, "Annual Energy Outlook 2013," U.S. Energy Information Administration2013.

[4] C. T. Hendrickson, L. B. Lave, and H. S. Matthews, *Environmental Life Cycle Assessment of Goods and Services: An Input-Output Approach*: RFF Press, 2006.

[5] Y. Chang, R. J. Ries, and Y. Wang, "The quantification of the embodied impacts of construction projects on energy, environment, and society based on I–O LCA," *Energy Policy,* vol. 39, pp. 6321-6330, 2011.

**(387d) Root-Cause Analysis in the Design of Chemical Processes**

Gangadharan, P., Lamar University
Chen, L., Lamar University
Lou, H. H., Lamar University
Originally presented on: 11/5/2013 16:12:00 - 16:31:00

Root-Cause Analysis techniques have found widespread application in quality control, management, failure analysis and process safety. A sustainability root-cause analysis is a powerful tool that will enable designers to identify and display the fundamental causes, related causes and root causes that impact the sustainability of a process. This facilitates the understanding of potential interrelationships between causes which lead to the sustainability issue. A base case in the design of chemical/energy systems generally has high potential for improvement. The challenge is to identify the bottleneck issues that can have a significant impact on overall sustainability.

The Pareto chart and the Fish-bone diagram, in conjunction with the sustainability metrics, form the framework of the sustainability root-cause analysis. The sustainability assessment incorporates economic, environmental, societal, and efficiency criteria. Because of the multi-dimensional feature of sustainability, accounting for the impacts of various design factors and the cause-and-effect relationships can be very difficult. After selecting a process from a group of alternative designs, it is further examined in a root-cause analysis, to identify the major causes that affect the parameters considered in the sustainability assessment, and then eliminate them with appropriate corrective action.

To illustrate the use of the sustainability root-cause analysis tools, several case studies including biodiesel production, polygeneration and natural gas reforming are employed. The Pareto analysis and Fish-bone diagram are invaluable aids to visualize complex relationships and screen out significant factors and correlations for sustainability improvement. This methodology can provide critical guidance to design for sustainability.

**(387e) Development of Product-Based Energy Intensity Metrics for the California Food Industry**

Walker, M. E., Illinois Institute of Technology
Hutchins, M., UC-Berkeley
Lettieri, D., UC-Berkeley
Romanin, V., UC-Berkeley
Blinde, P.,
Masanet, E., Northwestern University
Originally presented on: 11/5/2013 16:31:00 - 16:50:00

Product-based energy intensity metrics describe the energy footprint of manufactured products. These metrics are valuable for several reasons. From an industry standpoint, determining these metrics are an important first step to the development of an internal energy benchmarking strategy. From a marketing point of view, industrial and home consumers alike are putting an increased focus on green manufacturing and carbon footprints; companies that can highlight their own energy and manufacturing efficiencies may benefit in the marketplace. Finally, in terms of emissions regulation, the development of product-based energy metrics and a product-based benchmarking strategy is the preferred approach.

This presentation focuses on the development of product energy intensity metrics in the California food industry for integration with the state's Cap & Trade program. The development of such metrics for food processing presents a number of challenges, stemming primarily from the fact these plants manufacture multiple products and may lack detailed process monitoring capabilities. This presentation highlights the issues facing the determination of these metrics and discusses the strategies and methodologies applied to develop these metrics in the CA food processing industry.

**(387f) Estimations of Direct and Indirect Emissions From Mobile Air Conditioner Units in Automobile Transportation: The Cool Mile Functional Unit**

Blowers, P., The University of Arizona
Alayedi, I.,
Dougless, A., U of Arizona
Gerwe, B., U of Arizona
Mesta, A., U of Arizona
Rugel, C., U of Arizona
Wang, Y. Y.,
Originally presented on: 11/5/2013 16:50:00 - 17:09:00

Hydrofluorocarbons (HFCs) may become regulated in the near future by the US Environmental Protection Agency because of their contributions to global climate change. HFCs are currently used as refrigerants and solvents because they are relatively inexpensive and inert. This work focuses on the use of HFCs in automobile air conditioning applications because of the large number of vehicles in use in the US, and the rapid rise of the application of refrigeration in mobile air conditioning systems (MACs) worldwide, with R134a being the predominantly used refrigerant. R-134a (1,1,1,2-tetrafluoroethane,) has emissions of about 38.6 x $10^6$ kg/year per year from all sources, with approximately 47% of those due to emissions from vehicles. If R-134a is banned, replacements need to be evaluated in order to determine if greenhouse gas (GHG) emissions are indeed reduced when both direct emissions, in the form of leaks, and indirect ones from changes in fuel mileage, are accounted for.

Papasavva, et al, hafe already analyzed emissions patterns for a new replacement compound, HFc-1234hf, which is a hydrofluoro-olefin from fleet-wide use of the refrigerant in different locations around the world under ambient weather conditions, but did not consider indirect emissions impacts. This refrigerant has also been technically evaluated for use in passenger cars and it appears to be a suitable drop-in replacement. Other refrigerant choices are possible and are investigated in this work, using the "number of cool-miles" as the functional unit in order to capture the changes in emissions due to air conditioning use during drawdown (initially cooling of a vehicle from a hot start) and/or stable maintenance of temperature on longer drives so that the importance of different variables can be explored.

This work shows that some refrigerants have reduced direct emissions due to the refrigerant having a lower global warming potential (GWP), but that indirect emissions due to running the refrigerant through the system sometimes outweigh those direct emissions reductions. Future work should explore how to capitalize on opportunities to minimize the sum of the direct and indirect emissions.

**(387g) Regional Air Quality Improvement By Scheduling of Multiple Chemical Plant Start-Ups**

Cai, T., Lamar University
Xu, Q., Lamar University
Originally presented on: 11/5/2013 17:09:00 - 17:28:00

Chemical plant concentrated regions may suffer localized and transient air pollution events that violate national ambient air quality standards (NAAQS). Flaring emissions, especially intensive start-up flaring emissions from chemical plants, have potentially significant impacts on local air quality. Thus, when multiple plants in an industrial zone plan to start-up within a same time period, their start-up plans should be evaluated and optimally controlled so as to avoid unexpected air-quality violations in any air-quality concern regions (AQCRs). In this paper, a general systematic methodology for multi-plant start-up emission evaluation and control has been developed. The methodology starts with collecting regional meteorological information such as wind speed and temperature; geographical information of all of the involved chemical plants and AQCRs; as well as plant operation data such as the start-up time window, start-up duration, and estimated emission profile. Next, a regional air-quality evaluation based on Gaussian dispersion model will be conducted. If any air quality violation is predicted to an AQCR, a multi-objective scheduling problem will be generated and solved to optimize the start-up sequence and start-up beginning time for all chemical plants. The scheduling model minimizes the overall air quality impacts to all of AQCRs as well as minimize the total start-up time mismatch of all plants, subject to the principles of atmospheric pollutant dispersion. This study may provide valuable quantitative decision supports for multiple stake holders, including government environmental agency, regional chemical plants, and local communities.

## (437a) Gas Purification and Separation With Ionic Liquids

Zhang, X., Institute of Process Engineering, Chinese Academy of Sciences
Dong, H., Institute of Process Engineering, Chinese Academy of Sciences
Huang, Y., Institute of Process Engineering, Chinese Academy of Sciences
Zhang, S., Institute of Process Engineering, Chinese Academy of Sciences
Zhang, X., Institute of Process Engineering, Chinese Academy of Sciences
Originally presented on: 11/6/2013 8:30:00 - 8:55:00

**Gas purification and separation with Ionic liquids**

Xiangping Zhang, Haifeng Dong, Ying Huang, Xin Zhang, Suojiang Zhang

Beijing Key Laboratory of Ionic Liquids Clean Process, State Key Laboratory of Multiphase Complex System, Institute of Process Engineering, Chinese Academy of Sciences, Beijing, 100190, China

Abstract

In the past decade, ionic liquids have emerged as a promising alternative to be used in the gas purification and separation due to their remarkable properties, such as negligible volatility, high chemical/physical stability, higher solubility and selectivity, and tenability which allows design of task-specific ionic liquids (TSILs). In industries, gas separation or processing was regarded as an important operation to remove harmful and toxic gas emissions from waste gas in order to reduce ecologic destroy and environmental pollution from an environmental perspective; on the other hand, purifying gases by separating impurities from mixtures can improve the quality of the gas to satisfy the requirement of gas products considering the industrial relevance. There are some typical gas separation processes with ionic liquids, such as $CO_2$ capture, desulfurization of flue gas, volatile organic compounds (VOCs) control, as well natural gas sweetening, have shown promising potentials for industrialization.

In this work, an Industrial Application-Oriented research framework was proposed and conducted to develop new processes to purify and separate gases taking ionic liquids as main solvents. The following studies have been focused on: 1) the structure-properties relationship of candidate ionic liquids was studied using the molecular dynamics simulation and the quantum mechanics, combining the results of spectrum characterization. Studying the interaction force between the gas and ionic liquid would be helpful to understand the absorption mechanism and then to determine the criterion of the designing and screening promising ionic liquids; 2) new thermodynamic models were developed and used to predict the physicochemical properties of ILs on the base of the fragment contribution-corresponding states (FCCS) method. The FCCS method has advantages to deal with both the molecular and ionic systems, and it is more precise than traditional method with an average absolute relative deviation less than 4% for more than 600 kinds of ILs investigated. The binary and ternary-compound gas-liquid phase equilibrium data were measured in the lab, and the SRK equation of state was used for predicting the solubility gas in ILs. The binary parameters for VLE of compounds were regressed from experimental data; 3) the transportation properties of gas and liquid were important for reactor design. The multi-bubble/single behavior in ionic liquids was studied by using high speed image pick-up system. The bubble behavior, such as bubble size distribution, gas holdup, Sauter diameter and interfacial area were investigated. New correlations based on the experimental data was proposed for the prediction of bubble Sauter diameter in gas-ionic liquid systems; 4) process simulation was used to determine the operational parameters and the flowsheet structure. The whole flowsheet of gas separation was established considering real industrial constraints. 5) techno-economic assessment procedure is also important to evaluate the new process. On the base of the models proposed and experimental data, the key parameters (e.g. solvent concentration and flow rate, stage numbers and lean loading) were optimized aiming at minimizing the energy requirement of the solvent regeneration. Capital costs and operating costs were estimated in this work to assess the economic performance.

Several ionic liquid-based gas separation processes were studies on the base of the above research framework. The possibilities and the feasibilities of the new processes would be expected to be obtained. The relevant study not only provides new gas separation technologies to replace the traditional energy-extensive and polluted ones, but also indicates the advantages and drawbacks before these new processes come into commercialization. Importantly, the obtained information will be helpful to avoid the failures of industrialization and reduce the risk of investment in the future.

**Acknowledgements**

We would like to acknowledge the support from Key Program of National Natural Science Foundation of China (No.21036007), the National Natural Science Foundation of China (No. 51274183) and the National High Technology Research and Development Program of China (No.2011AA050606).

**(437b) Are Ionic Liquids â€œDrop-Inâ€ Replacements for Conventional CO2 Capture Solvents in Packed Absorption Towers? Capital Considerations**

Minnick, D. L., University of Kansas
Weatherley, L. R., University of Kansas
Scurto, A. M., University of Kansas
Originally presented on: 11/6/2013 8:55:00 - 9:20:00

Ionic liquids are actively being pursued as absorption solvents for $CO_2$ capture. Some ionic liquids have been found to have some of the largest $CO_2$ absorption capabilities of any known solvent. However, the relatively slower mass transport properties of ionic liquids have been mentioned as potential challenges to their industrial implementation. This presentation will illustrate the performance of the known and hypothetical ionic liquids for their use in absorption towers. The size and hydraulics of absorption towers using ionic liquids will be compared with conventional alcohol-amine solvents. The interplay of the thermodynamic solubility with the mass transfer rates has large effects on the dimensions and hydraulics (capital costs) of packed absorption towers. Our findings suggest that even ionic liquids with the highest known $CO_2$ capacities can yield capital costs in excess to conventional solvents due to the packed tower specifications necessary to overcome their slower mass transfer rates. Generally, improvements in the mass transport properties often yield smaller absorption towers compared to improvement in the thermodynamic $CO_2$ solubility. This may indicate that other process intensification techniques may need to be considered to take advantage of some of the real thermodynamic advantages of ionic liquids. Molecular design parameters are suggested for the improvement in ionic liquids for $CO_2$-capture.

**(437c) Post-Combustion CO2 Capture With Ionic Liquids Involving Phase Change**

Seo, S., The University of Notre Dame
Simoni, L. D., University of Notre Dame
DeSilva, A., University of Notre Dame
Massel, M., The University of Notre Dame
Stadtherr, M. A., University of Notre Dame
Brennecke, J. F., University of Notre Dame
Originally presented on: 11/6/2013 9:20:00 - 9:40:00

Ionic liquids (ILs) are organic salts that are liquids below the boiling point of water. Non-volatility, good thermal stability and high carbon dioxide ($CO_2$) solubility make ILs an attractive replacement for commercially available volatile $CO_2$ absorbents. A new concept for $CO_2$ capture that uses ***phase change ionic liquids (PCILs)*** offers the potential to significantly reduce parasitic energy losses incurred from capturing $CO_2$ from flue gas. PCILs are solid ionic materials that react stoichiometrically and reversibly with $CO_2$ (one mole of $CO_2$ for every mole of salt at typical post-combustion flue gas conditions) and form a liquid when they react with $CO_2$. This allows for a novel process that uses the heat of fusion ($\Delta H_{fus}$) to provide part of the heat needed to release $CO_2$ from the absorbent, reducing the total energy required. In addition, the phase transition yields almost a step-change absorption isotherm, which would require smaller pressure/temperature swing in the stripper. We present here detailed measurements of the $CO_2$ uptake, heat of absorption ($\Delta H_{abs}$) and phase transition behavior for selected PCIL compounds as well as the development of an absorption isotherm model to capture the underlying thermodynamics of the $CO_2$ absorption and phase change process. Additionally, the performance of lab scale absorption unit and regeneration unit with a PCIL will be discussed.

**(437d) Bubble Behavior in Seperation Process With Ionic Liquids**

Dong, H., Institute of Process Engineering, Chinese Academy of Sciences
Zhang, X., Institute of Process Engineering, Chinese Academy of Sciences
Bao, D., Institute of Process Engineering, Chinese Academy of Sciences
Zhang, X., Institute of Process Engineering, Chinese Academy of Sciences
Zhang, S., Institute of Process Engineering, Chinese Academy of Sciences
Originally presented on: 11/6/2013 9:40:00 - 10:00:00

**(437e) Permeability of Carbon Dioxide Through Ionic Liquid Membrane With Polyvinylidene Fluoride**

Shimoyama, Y., Tokyo Institute of Technology
Komuro, S., Tokyo Institute of Technology
Jindaratsamee, P., Tokyo Institute of Technology
Originally presented on: 11/6/2013 10:00:00 - 10:20:00

Membrane separation is expected to be candidate for carbon dioxide capture. Several materials, such as organic polymer, ceramics and hybrid of organic – inorganic, have been used for the membrane separation of carbon dioxide. A supported liquid membrane is a porous membrane filled with liquid. For carbon dioxide separation, amine solvents are often used as the liquid for the liquid membrane, but there is a problem of the low membrane stability due to the evaporation of the solvent. Ionic liquid can be suitable for the liquid membrane solvent more than the amines because their vapor pressure is ultra low. It is important for the process design of the membrane separation to understand the permeability of carbon dioxide through ionic liquid membrane. In this work, the permeabilities of carbon dioxide through ionic liquid membranes were investigated. Polyvinylidene fluoride (PVDF) was used as the porous membrane. Six species of ionic liquid were used as follows: 1-ethyl-3-methylimidazolium tetrafluoroborate ([emim][BF4]), 1-butyl-3-methylimidazolium bis(trifluoromethylsulfonyl) imide ([bmim][Tf2N]), 1-butyl-3-methylimidazolium tetrafluoroborate ([bmim][BF4]), 1-butyl-3-methylimidazolium hexafluorophosphate ([bmim][PF6]), 1-butyl-3-methylimidazolium trifluoromethanesulfonate ([bmim][OTf]) and 1-butyl-3-methylimidazolium dicyanamide ([bmim][dca]). The carbon dioxide permeabilities were measured by a sweep gas method at 303 to 343 K. Nitrogen gas was used as the sweep gas. The effects of temperature, anion species on carbon dioxide permeability though ionic liquid membrane were studied. The permeabilities of carbon dioxide increase with temperatures. The ionic liquid membrane with [bmim][Tf2N] results in the highest permeability of carbon dioxide of all the species in this work. The permeabilities of carbon dioxide through [bmim][PF6] and [bmim][Tf2N] membranes with water vapor were also measured at 303 to 333 K. It is found that the addition of water vapor at feed and permeate sides makes an enhancement of the permeabilities of carbon dioxide through ionic liquid membranes.

**(437f) Phase Equilibria and Separations Involved in the Synthesis of the Ionic Liquid 1-Hexyl-3-Methylimidazolium Bromide ([HMIm][Br]) in Acetone**

Minnick, D. L., University of Kansas
Scurto, A. M., University of Kansas
Originally presented on: 11/6/2013 10:20:00 - 10:40:00

A vast amount of research on ionic liquids appears in the open and patent literature. For the future industrial applications, large quantities of ionic liquids will be required. The scale-up of organic processes has been a longstanding mission of chemical engineering. However as ionic liquids are touted as potentially environmentally-benign solvents, ILs themselves must also be made in a sustainable manner. Our previous studies have measured the kinetics and initial human/environmental impact of the production of a model ionic liquid, 1-hexyl-3-methylimidazolium bromide ([HMIm][Br]) in a variety of solvents, including compressed $CO_2$. The conventional solvent acetone was determined to be a leading convetional solvent both from a reaction engineering standpoint but also an environmental. Here, the experimental vapor-liquid equilibrium and liquid-liquid equilibrium involved in the synthesis mixtures of [HMIm][Br] from 1-bromohexane and 1-methylimidazole in acetone have been performed. The experimental data is modeled using both the Peng-Robinson equation of state and excess Gibbs/activity coefficient models. These models are then used to simulate potential separation methods with particular attention on the energy demands on producing the ionic liquid at a given purity.

**(437g) A Promising Electrolyte of Ionic Liquids for Aluminium Deposition**
Wang, Q., Institute of Process Engineering, Chinese Academy of Sciences
Originally presented on: 11/6/2013 10:40:00 - 11:00:00

A Promising Electrolyte of Ionic Liquids for Aluminium Deposition

Qian Wang[a], Bin Chen[a], Qinqin Zhang[a,b], Suojiang Zhang[a,*]

[a]Beijing Key Laboratory of Ionic Liquids Clean Process, Key Laboratory of Green Process and Engineering,

State Key Laboratory of Multiphase Complex Systems, Institute of Process Engineering, Chinese Academy of Sciences,

Beijing 100190, P.R. China

[b]College of Chemistry and Chemical Engineering, University of Chinese Academy of Sciences, Beijing 100049, P.R. China

[*]Corresponding author. E-mail: sjzhang@home.ipe.ac.cn Tel.: +86 10 82627080

Ionic liquids (ILs) become the preferred electrolyte for aluminum electrodeposition due to their low melting point, negligible vapor pressure, high conductivity, and wide electrochemical window [1]. The research work on electrodeposition of aluminum from ILs began in 1951 [2], since then researchers conducted a series of studies [3-5], inwhich the types of ILs can be divided into four categories. They are mixtures of anhydrous aluminium halides with alkylpyridinium halides, alkylimidazolium halides, alkylpyrrolidinium halides and aralkylammonium halides, respectively. Usually, the anhydrous aluminium halide is anhydrous aluminium chloride ($AlCl_3$).

In this study, four typical ILs of the above four systems, 1- butyl -3- methylpyridinium chloride-$AlCl_3$, 1-butyl-3- methylimidazolium chloride-$AlCl_3$, 1-butyl-1-methylpyrrolidinium chloride-$AlCl_3$, and phenyltrimethylammonium chloride-$AlCl_3$, were chose using as the electrolyte for aluminium deposition. Firstly, the physical properties of the four ILs were measured. At room temperature, 1-butyl-3-methylimidazolium chloride-$AlCl_3$ has the highest conductivity and the lowest viscosity. Meanwhile, they are similar in density. Then aluminium coatings were prepared from them at the same current density, respectively. By comparing the quality of the obtained coatings, it was found that compact and smooth aluminium coatings can be prepared from the IL formed by mixing 1-butyl-3- methylimidazolium chloride with $AlCl_3$ at 30℃. But in the other three systems, only when the temperature is higher than 60℃, good quality aluminium coatings can be obtained. Finally, the cyclic voltammetry of the four electrolytes was determined in a three-electrode cell at different temperature, also showing that the 1-butyl-3-methylimidazolium chloride-$AlCl_3$ has the best electrochemical property at a lower temperature. All this proves that 1-butyl-3- methylimidazolium chloride-$AlCl_3$ is a promising electrolyte for aluminium deposition.

**References**


[1] Brown L. Electroplating with non-aqueous solutions [J]. *Trans. Inst. Met. Finish.*, 2010, 88,122-123.

[2] Hurley F H, Wier T P. The electrodeposition of aluminum from nonaqueous solution at room temperature [J]. *J.Electrochem.Soc.*, 1951, 98, 207-212.

[3] Yue G, Zhang S, Zhu Y, et al. A promising method for electrodeposition of aluminium on stainless steel in ionic liquid [J]. *AIChE J.*, 2009, 55, 783-796.

[4] Liu Q, Liu K, Han Q, et al. Study on electrodeposition of Al from trimethylphenylammonium chloride (TMPAC)-aluminum chloride ($AlCl_3$) ionic liquids + benzene. *Chin. Surf. Eng.*, 2010, 23, 34-39.

[5] Giridhar P, El Abedin S Zein, Endres F. Electrodeposition of aluminium from 1-butyl-1-methylpyrrolidinium chloride/$AlCl_3$ and mixtures with 1-ethyl-3-methylimidazolium chloride/$AlCl_3$ [J]. *Electrochim. Acta*, 2012, 70, 210-214.

**(444a) Novel Catalysts and Reactor Concepts for Sustainable Processes: the Use of LCA Data in Process Design**

Yaseneva, P.,
Lapkin, A., University of Cambridge
Originally presented on: 11/6/2013 8:30:00 - 8:55:00

**Novel catalysts and reactor concepts for sustainable processes: the use of LCA data in process design**

**Polina Yaseneva and Alexei Lapkin**

*Department of Chemical Engineering and Biotechnology, University of Cambridge, Cambridge, CB2 3RA, United Kingdom*

## Abstract

We present a multi-step assessment framework, which was developed to guide decision making process in a process development context. Reaction mass intensity metric, gate-to-gate flowsheet analysis and LCA are used at appropriate stage gates. Specifically results of LCA assessment will be discussed for the case studies of a novel catalyst for water purification process, an intensive process of a pharmaceutical API synthesis and a novel organometallic catalytic process for intermediates synthesis.

## Introduction

Assessment of sustainability of novel processes for decision-making purposes is a complex task, as frequently uncertainty of the data precludes the use of full LCA methodology, whereas proxy methods such as mass or energy metrics provide incomplete information. At the same time, information about sustainability of novel technologies is now routinely requested even within research projects. Thus, most EU projects in the area of chemical technologies are accompanied by LCA studies.

Within a large integrated project SYNFLOW (www.synflow.eu) we have developed a three-step approach, using reaction mass efficiency metric for selection of chemical routes, gate-to-gate mass and energy metrics for assessment on flow sheet options and finally LCA for evaluation of the demonstration case studies. These steps coincide with stage-gating within a project: confidence in chemistry to be taken up for process development, evaluation of process options with a focus on clean technologies (novel solvents, process intensification) and finally the detailed analysis of complete process at the stage of demonstration with LCA, targeting detailed flowsheet optimisation.

Here we present results of LCA studies for three specific examples. We show LCA of a novel carbon nanotube-based catalyst for water purification, focusing on comparison of the manufacture of the catalyst, a study of a novel flow-process of stoichiometric reduction of artemisinin, leading to manufacture of an important pharmaceutical API (focusing on comparison of batch vs continuous process) and finally a study of a novel catalytic Buchwald-Hartwig amination under overall flow conditions.

**Results and Discussion**

*Carbon nanotube catalyst for water purification*

A highly active catalyst was developed for reduction of an emergent aqueous pollutant, bromate, based on 0.3 %wt Pd deposited onto carbon nanotubes, grown on sintered metal fibre. The TOF of the catalyst was found to be an order of magnitude higher than that of a conventional catalyst based on alumina support ($Pd/Al_2O_3$). However, there remained the question whether high-temperature CVD process used in the manufacture of carbon nanotubes resulted in a cleaner overall process. The LCA study focused on the manufacture of the two catalysts. In this specific case all environmental impacts for the new catalysts were lower. Detailed analysis of contributions of the individual manufacturing steps to various impacts, e.g. analysis of contributions to cumulative energy demand shown in Figure 1, allows to reveal further options for optimisation of the novel catalyst manufacture [1].

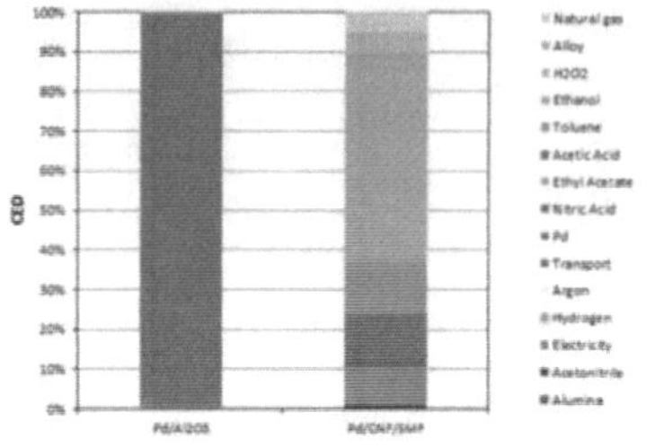

Figure 1. Individual contributions to cumulative energy demand (CED) of the manufacture of the two catalysts.

*Artemisinin reduction under intensive flow conditions*

We have developed a flow process for reduction of artemisinin to overcome the limitations of a conventional batch process [2]. In the current study we extended the reaction to the final step in the manufacture of the antimalarial API, but also performed an LCA study to compare the novel flow process with the conventional batch reaction [3]. Analysis of LCA results revealed that the solvent used in the manufacture and storage of the reducing agent required for the flow process is having a massive negative impact, see Figure 2. It is becoming critically important to replace the solvent used in a particular stage of the overall process. In this case LCA is instrumental in identifying the critical stage of the overall process that requires optimisation.

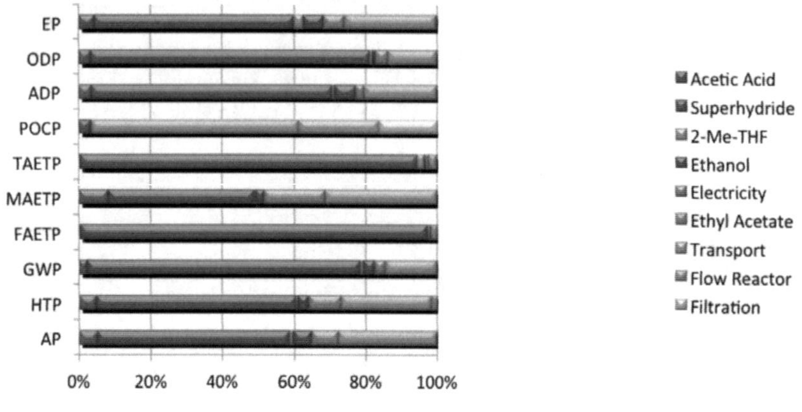


Figure 2. Individual contributions to environmental impacts in artemisinin reduction under flow conditions.

*Buchwald-Hartwig amination in flow*

The three-step decision making process is illustrated on the example of Buchwald-Hartwig amination. Reaction mass efficiency, process efficiency and LCA are evaluated for this reaction at the different stages in the process design. This approach complements the parallel chemistry-process design methodology being developed within Synflow project, ultimately aiming at a significant reduction of time required for commercialisation of novel technologies, which are also sustainable. Thus, for the initial reaction options we have used RME metric to evaluate the importance of solvent and catalyst recycle. The new process is currently being scaled-up to a demonstrator at an Invite facility of Bayer. We will show the results of stage-gating metrics and initial LCA data.

[1] P. Yaseneva et al., Reduction of bromates on carbon nanofibre supported structured Pd catalysts: experimental and life cycle assessment study, Submitted.

[2] X. Fan, V. Sans, P. Yaseneva, D. Plaza, J.M.J. Williams, A. Lapkin, Facile stoichiometric reductions in flow:

example of artemisinin, Org Proc Res Dev 16 (2012) 1039-1042.

[3] P. Yaseneva et al., Optimisation of the flow process of manufacture of artemether and its LCA study, Submitted.

**Acknowledgement:** This work was supported by the European Community's Seventh Framework Programme under project "SYNFLOW", NMP2-LA-2010-246461.

**(444b) Techno-Economic and Environmental Assessment of Ionic Liquid-Based Biorefineries**

Giarola, S., Imperial College London
Shah, N., Imperial College London
Originally presented on: 11/6/2013 8:55:00 - 9:20:00

Although the peak oil is still a debated issue, petroleum supply is expected to become ever more vulnerable of interruptions and current economic system reliance on fossils is exposing all economic sectors to unsustainable price fluctuations [1]. Concerns of maintaining an adequate supply of resources, such as fuels, food, raw materials for a growing population, are also raising a pressing interest for a change in our way to conceive energy and resource provision system [2]. A paradigm shift in the industrial sector from today's large fossils exploitation towards more sustainable practices is also advocated to face current trend of carbon dioxide emissions production, which may soon be approaching greenhouse gas (GHG) levels dangerously close to tipping points of an accelerated global warming [3]. Thus, the future challenge to address the growing world energy demand may be met as long as the supply of higher levels of resources and energy is achieved in a sustainable way promoting low-carbon technologies and renewables [4]. In this respect, the spread of biomass exploitation, as a unique source that can supply both energy and carbon for the production of the commodity chemicals and products, has the potentials to lead to the settlement of a new biomass-based economy [5-6].

Biomass is an abundant, renewable and carbon neutral resource, which is currently significantly exploited for biofuels production from edible components of food crops. However, the major challenge for a future bio-based economy would very likely rely on lignocellulosic feedstock exploitation, being more abundant and less expensive than starchy crops as well as avoiding any direct conflicts with the food chain [7]. Industrial biorefineries have been identified as the most promising route to the creation of a new domestic biobased industry, where biomass conversion processes into fuels, power, and chemicals are integrated through an efficient combination of technologies, to minimise the environmental footprints and ensure the sustainability of all products generated. By producing multiple products, a biorefinery can take advantage of the differences in biomass components and intermediates and maximise the value derived from the biomass feedstock [8]. However, cost-effective technologies for processing lignocellulosic materials into fuels and chemicals are not commercially available. In this context, a central role in determining the overall system profitability is played by the pretreatment unit, where biomass is deconstructed and the three main components (i.e. cellulose, hemicellulose and lignin) are made accessible to subsequent processes [9].

Since the study of Wilkes and Zaworotko [10], ionic liquids have become increasingly popular as reaction and extraction media. The property of ionic liquids which has led to them being classed as green solvents is their very low volatility, thus they do not emit volatile organic compounds and are suitable for reactions at atmospheric pressure. Recently, ionic liquids have begun to be investigated for biomass related processes, yet there are many technological applications and large commercial interests in the field [2]. The use of ionic liquids as lignocellulose deconstruction solvents is also extremely promising [11]. Among the potential applications, ionic liquid can be used to decrystallise cellulose and disrupt the lignin-hemicellulose network. The possibility of removing lignin with ionic liquids and recovering it as precursor for commercial products is also an interesting option [11]. However, deconstruction with ionic liquid will only be viable if some drawbacks are overcome, in particular the cost of ionic liquids. Process design and optimisation is required in order to have high solvent recovery and recycle, but also consumables use reduction, energy integration and waste minimisation. Economic aspects should also be coupled with the concept of ecological sustainability during the process optimisation strategy [12]. Process system engineering techniques play a major role in achieving all the above targets.

This work proposes an investigation of techno-economic performance of a biorefinery based on ionic liquid-pretreatment. The analysis, based on mathematical modelling and process design techniques, aims at identifying the key physical phenomena, the main cost centres affecting the process economics and proposing strategies to develop more efficient biochemical conversion routes for the biomass. This is performed along with the system environmental performance appraisal based on Life Cycle Assessment (LCA) techniques [13]. The final purpose is to provide a mathematical tool for an in-depth economic and environmental analysis of the ionic liquids fractionation, identifying the process enhancements alternatives and outlining potential synthetic routes to biocommodities.

*References:*


1. Lutz, C., U. Lehr, K. S. Wiebe (2012). Economic effects of peak oil. Energy Policy, 48, 829-834.
2. Tan, S. S. Y., D. R. MacFarlane (2009). Ionic liquids in Biomass Processing. Topics in current chemistry, 290, 311-340.
3. Foley, J. A. (2005). Tipping Points in the Tundra. Science, 310, 627-628.
4. International Energy Agency (2012). World Energy outlook. IEA, Paris (FR).
5. Clark, J. H., F. E. I. Deswarte, editors (2008). Introduction to Chemicals from Biomass Padstow: John Wiley & Sons, Ltd.
6. Ragauskas, A. J., C. K. Williams, B. H. Davison, G. Britovsek, J. Cairney, C. A. Eckert, et al. (2006). The Path Forward for Biofuels and Biomaterials. Science, 311, 484-489.
7. The Royal Society (2008). Sustainable biofuels: prospects and challenges. The Royal Society, London (UK). www.royalsociety.org
8. National Renewable Energy Laboratories (2009). Biomass Program. U.S. Department of Energy, Golden (CO).
9. Aden, A., M. Ruth, K. Ibsen, J. Jechura, K. Neeves, J. Sheehan, et al. (2002). Lignocellulosic biomass to ethanol process design and economics utilizing co-current dilute acid prehydrolysis and enzymatic hydrolysis for corn stover. Technical Report. National Renewable Energy Laboratory (NREL), Golden (CO).
10. Wilkes, J. S., M. J. Zaworotko (1992). Air and water stable 1-ethyl-3methylimidazolium based ionic liquids. Journal of the Chemical Society, Chemical Communications, 965-967.
11. Brandt, A., J. Gräsvik, J. P. Hallett, T. Welton (2013). Deconstruction of lignocellulosic biomass with ionic liquids. Green Chemistry, 15, 550-583.
12. Kralisch, D., D. Reinhardt, G. Kreisel (2007). Implementing objectives of sustainability into ionic liquids research development. Green Chemistry, 9, 1308-1318.
13. International Organisation for Standardisation (1997). ISO 14040: Environmental Management - Life Cycle Assessment - Principles and Framework. International Organisation for Standardisation: (ISO), Geneva.

**(444c) Optimal Simultaneous Production of Bio-Ibutene and Bioethanol From Switchgrass**
MartÃn, M., University of Salamanca
Grossmann, I. E., Carnegie Mellon University
Originally presented on: 11/6/2013 9:20:00 - 9:45:00

Ibutene is an important intermediate that has been typically produced from the cracking of the C4 fraction of crude oil. It is the basis for the production of a common additive to gasoline in search for a cleaner burning fuel, the methyl-tert-butyl ether (MTBE), as well as a monomer for the polymerization such as butyl rubber. Its importance can be somehow reflected in its price, around $2/kg. However, the dependency on the crude oil and the already limited availability due to the large number of applications has increased the need for new sources of this chemical. Ibutene has recently been used as intermediate for the production of oxygenated diesel substitutes from glycerol (Chen et al 2011, Vlad et al 2010, Martín & Grossmann 2013). In this application the main drawbacks are, on the one hand, the production of diesel substitutes using a fossil fuel based intermediate and, on the other hand, the economics since from the energy and water consumption standpoint its use for the enhanced production of diesel substitutives is competitive with the process that directly sells the glycerol as byproduct.

In this work, we propose the optimal flowsheet for the production of ibutene from switchgrass. A superstructure embedding a number of alternatives is proposed. Two technologies are considered for switchgrass pretreatment, dilute acid and ammonia fibre explosion (AFEX) so that the structure of the grass is broken down. Surface response models are developed for predicting the effect of the operating conditions on the yield of each pretreatment. Next, enzymatic hydrolysis follows any of the pretreaments to obtain fermentable sugars, mainly xylose and glucose. Ibutene is obtained by fermentation of the sugars. However xylose cannot be easily converted, and thus we also evaluate the possibility of using it to produce bioethanol. Next the ibutene has to be separated mainly from $CO_2$ for which, PSA or membrane separation are considered while in case ethanol is produced, a multi-effect distillation column followed by molecular sieves are used to dehydrate the ethanol. The problem is formulated as an MINLP with simultaneous optimization and heat integration based on Duran and Grossmann's model (1986). Finally, an economic evaluation is performed.

The most promising process involves the use of dilute acid pretreatment and membrane purification of the ibutene. The production of ibutene using renewable sources is feasible, and may be competitive for its use to generate diesel substitutes for glycerol etherification. However, the decision related to the production of ibutene alone or the simultaneous production of ibutene and ethanol depends on the prices for bioethanol and for switchgrass. For low and current biomass cost the production of ethanol ad ibutene is interesting for bioethanol process above $1/gal.

References:

Chen, P., Min, M., Chen, Y., Wang, L., Li, Y., Chen, Q., Wang, C., Wan, Y., Wang, X., Cheng, Y., Deng, S., Hennessy, K., Lin, X., Liu, Y., Wang, Y., Martinez, B., Ruan, R., 2009. Review of the biological and engineering aspects of algae to fuels approach. Int. J. Agric. Biol. Eng. 2, 1–30.

Duran, M.A.; Grossmann, I.E. (1986) Simultaneous optimization and heat integration of chemical processes. *AIChE, J.*, 32, 123-138

Martín, M., Grossmann, I.E. (2013) Simultaneous dynamic optimization and heat integration for the co-production of diesel substitutes: Biodiesel (FAME & FAEE) and glycerol ethers from algae oil. To be Submitted

Vlad, E., Bildea, C.S. and Bozga, G., 2010, Integrated design and control of glycerol etherification processes, Bull. Inst. Pol. Iasi, LVI (LX), 4, 139-148.

## (444d) Synthesis and Design of Biorefinery Processing Networks With Uncertainty and Sustainability Analysis

Cheali, P., Technical University of Denmark
Gernaey, K. V., Technical University of Denmark (DTU)
Sin, G., Technical University of Denmark
Originally presented on: 11/6/2013 9:45:00 - 10:10:00

**Synthesis and Design of Biorefinery Processing Networks with Uncertainty and Sustainability analysis**

**Peam Cheali, Krist V. Gernaey and Gürkan Sin**

Department of Chemical and Biochemical Engineering, Søltofts Plads, Building 229, Technical University of Denmark, DK-2800 Lyngby, Denmark

Chemical industries usually rely on fossil based feedstock, which is a limited resource. In view of increasing energy demands and the negative environmental and climate effects related to the use of fossil based fuels, this motivates the development of new and more sustainable technologies for processing renewable feedstocks, with the aim of bridging the gap for fuel, chemical and material production. This project is focusing on biorefinery network design, in particular for early stage design and development studies.

Optimal biorefinery design is a challenging problem. It is a multi-objective decision-making problem not only with respect to technical and economic feasibility but also with respect to environmental impacts, sustainability constraints and limited availability & uncertainties of input data at the early design stage. It is therefore useful to develop a systematic methodology to support the evaluation of processes and the generation of sustainable alternatives for identifying the optimal processing routes. One particular challenge here is to include proactively sustainability analysis as part of the synthesis of biorefinery networks. Another challenge is the handling of several sources of uncertainties such as availability and composition of renewable feedstock, the technical performance of alternative processing technologies and the availability of future markets for biorefinery products, among others.

As part of earlier work in our research group, a systematic methodology to identify/generate optimal biorefineries was developed using the superstructure-based approach, and was implemented in a computer-aided framework. The methodology consists of tools and methods including databases, models, superstructure, and solution strategies to represent, describe and evaluate various combinations of processing networks. The optimization of the network is formulated as a mixed integer nonlinear programming type of problem and solved in GAMS. The methodology was applied for designing optimal biorefinery networks considering biochemical routes. Furthermore, the methodology has also been applied to soybean oil processing networks and industrial wastewater treatment networks. In the work presented here, the methodology for designing optimal biorefinery networks was expanded with inclusion of thermochemical routes, which means the scope and the size of the biorefinery network problem was extended at the level of the database, the models and the superstructure relevant for thermo-chemical conversion routes (the conversion of biomass feedstocks (corn stover, poplar wood) to fuels and chemicals (FT-gasoline, FT-diesel, bioethanol and higher alcohols) via thermal decomposition processes like pyrolysis and gasification).,

In this study, we extend the methodology, models and database by incorporating uncertainty and sustainability analysis as well. Appropriate ranges for uncertain parameters are identified with their correlation/covariance structure and latin hypercube sampling (LHS) is used to sample parameters from their respective domain of uncertainty. The parameter samples are then used as input for solving a deterministic and stochastic optimization problem. The sustainability analysis was performed in two ways: First, it was performed retrospectively to the optimal biorefinery network solution obtained after the MINLP by using an in-house software (SustainPRO) that employs ICHEME sustainability metrics. Secondly, the sustainability analysis was included proactively as part of the MINLP optimization problem that is performed to find the optimal processing path with respect to multi-criteria assessment including technical, economics and sustainability.

The expanded database and superstructure with uncertainty and sustainability analysis form a powerful process synthesis toolbox to be used in design of future biorefineries with multi-criteria evaluation (technical and economic feasibility, environmental impact and sustainability).

**(444e) Ecosystems As Unit Operations for Innovative and Sustainable Process Design**

Bakshi, B. R., The Ohio State University
Urban, R. A., The Ohio State University
Ziv, G.,

Originally presented on: 11/6/2013 10:10:00 - 10:35:00

Since their development in the 1920s, the concept of unit operations has played an important role in the design of chemical processes through the development of general models and software tools. With increasing importance of reducing the environmental impact of chemical processes, the concept of unit operations expanded to include processes for reducing pollutants in the streams leaving the process boundary. Systems engineering methods based on pinch analysis and optimization have also played an important role in enhancing the efficiency of manufacturing systems. For most chemical processes, it is not practical or economically feasible to completely eliminate material emissions or waste, and the chemical industry continues to be a major producer of hazardous waste and air pollutants, including carbon dioxide.

Ecosystems have the natural ability to treat a large variety of pollutants, and provide goods and services to manufacturing systems. For example, wetlands can provide fresh water by removing contamination by chemicals such as pesticides, oils, phenolic compounds, nitrates, etc. Trees provide water, biomass and oxygen, along with the ability to remove air pollutants such as carbon monoxide, particulate matter, sulfur oxides, carbon dioxide, etc. In fact, an unsustainable system may be defined as a system that produces waste in excess of the natural cleaning capacity of ecosystems. Conversely, sustainability requires that engineering be done within ecological constraints. Unfortunately, most current sustainable engineering methods, including life cycle assessment are not able to satisfy this requirement.

In this presentation, we will "rediscover" the bounty of nature and its importance in sustaining chemical processes. This will involve expanding the concept of unit operations by including ecological processes, to enable engineering design within ecological constraints. Process flowsheets will be designed as networks of technological and ecological systems. Monetary and environmental objectives will be considered, and results for objectives such as maximum profit, zero water, and zero carbon dioxide will be compared. Application to a biodiesel manufacturing process will show that the techno-ecological network is superior to a conventional technology-only process in terms of economic and environmental objectives. Case studies of other chemical processes will demonstrate the need to modify the chemistry or processing methods to permit full benefit of ecosystems. This can lead to new innovation that may not be discovered without including ecosystems among the design options. This approach of techno-ecological networks may be extended to larger spatial scales to result in unique ways of designing and operating supply chains and entire life cycles so that they stay within ecological constraints.

**(444f) Process Analysis of Continuous Catalytic Gasification As a Waste to Energy Alternative**

Tillie, C. F., Cleveland State University
Riedthaler, M. J., Cleveland State University
Adhikari, S., Cleveland State University
Reeves, S. A., Cleveland State University
Gatica, J. E., Cleveland State University
Originally presented on: 11/6/2013 10:35:00 - 11:00:00

Presently, space exploration waste is either carried onboard for the duration of the mission or discarded into a trash module which is sent back into Earth's atmosphere for destruction. On the other hand, with an estimated cost of $10,000 per pound, the cost of delivering mass to locations beyond low earth orbit (LEO) locations is prohibitive.

Since approximately 15 pounds of waste are produced daily by a crew of four during a mission, the intrinsic value of this waste can be greatly enhanced by waste to energy technologies. Moreover, conceptualizing waste gasification as an in-situ alternative for processing of trash is also an efficient route to control waste, while maintaining a healthy habitable environment, during long-duration missions (Santiago-Maldonado et al., 2010).

The gasification of long-chain polymers is hypothesized to follow a mechanism of four combined reactions —two parallel oxidation reactions which are kinetically controlled, accompanied by two series reaction where oxidation products (carbon dioxide and carbon monoxide) are converted into methane and hydrogen. Integrating this process into spaceflight missions would greatly reduce the onboard mass requirements, as these products can be used as propellant fuel. Further, under certain conditions the process could be also be adapted to produce oxygen and water, two essential life support species for space missions.

This paper explores scale-up alternatives to implement low-temperature continuous gasification processes as potential waste management processes. A kinetic model formulated from batch gasification experiments is integrated into a process simulator along with a detailed thermodynamic description.

Emphasis is placed on the preparatory treatment of the waste slurries, the reactor module, and equilibrium stages for recycling reactants and recovering the catalyst. Detailed characterization of low and mid fidelity simulants is presented.

**(445a) Ethyl Acetate Synthesis Using Semi-Batch Reactive Distillation**

Kao, Y. L., National Taiwan University
Hsu, L. W.,
Ward, J. D., National Taiwan University
Originally presented on: 11/6/2013 8:30:00 - 8:51:00

Batch reactive distillation (BRD) is an integrated process which combines the advantages of reactive distillation and the flexibility of batch processes. Semi-batch reactive distillation (SBRD) is a design alternative for BRD in which one reactant is fed continuously into the column during the batch. Esterifications of low molecular weight alcohols (methanol–amyl alcohol) with acetic acid have been studied as model systems for the design and optimization of BRD and SBRD processes.

For such esterification processes the lowest-boiling stationary point is an azeotrope. If the azetrope is located in a two-liquid region and the aqueous phase is nearly pure water, (*n*-butanol and amyl alcohol), a BRD process is feasible because water (a product) can be removed from the column by a decanter.[1] Furthermore, Qi and Malone[2] showed that for the more difficult isopropyl acetate system in which the ternary azeotrope lies just inside the liquid-liquid zone, semi-batch reactive distillation (SBRD) operation can improve the production efficiency. Using acetic acid as an entrainer and continuously feeding it into the column during the batch can reduce the concentration of isopropyl alcohol in the distillate so that the distillate composition is closer to the water-acetate edge. Therefore the concentration of water in the aqueous phase is higher.

In this work, the more challenging ethyl acetate process is investigated. Unlike the isopropyl acetate process, BRD with a decanter is not feasible for the ethyl acetate process because the minimum-boiling ternary azeotrope is not located in the two-liquid region. To overcome this problem, researchers have considered the use of a middle-vessel column[3] and a two-column pressure swing BRD process.[4] In this work, we consider another possibility, which is the application of SBRD. Similar to the isopropyl acetate process, acetic acid is fed continuously near the top of the column to reduce the ethanol concentration in the distillate and move the distillate composition into the two liquid-liquid region. After the liquid split, high purity water can removed from the column, and the organic phase containing mainly ethyl acetate and water is collected. High purity of ethyl acetate (99 mol %) can be obtained from the separation of the collected organic product using a batch stripper with a decanter. The two step process is discussed in this work, and some important design variables including the number of reactive trays, initial amount of acetic acid added, side feed policy, reflux ratio policy, and reboiler heating policy are studied to improve the production efficiency.

## References

1. Venimadhavan G, Malone MF, Doherty MF. A novel distillate policy for batch reactive distillation with application to the production of butyl acetate. *Ind Eng Chem Res.* Mar 1999;38(3):714-722.
2. Qi W, Malone MF. Semibatch Reactive Distillation for Isopropyl Acetate Synthesis. *Ind Eng Chem Res.* Feb 2 2011;50(3):1272-1277.
3. Steger C, Lukacs T, Rev E, Meyer M, Lelkes Z. A Generic Feasibility Study of Batch Reactive Distillation in Hybrid Configurations. *Aiche Journal.* May 2009;55(5):1185-1199.
4. Modla G. Reactive pressure swing batch distillation by a new double column system. *Computers & Chemical Engineering.* Nov 15 2011;35(11):2401-2410.

**(445b) Coupled Permsmbr - Process Design and Development for 1,1-Dibutoxyethane Production**

Pereira, C. S., Laboratory of Separation and Reaction Engineering, University of Porto - Faculty of Engineering, Porto, Portugal
Silva, V. T., LSRE - Laboratory of Separation and Reaction Engineering - Faculty of Engineering - University of Porto
Rodrigues, A. E., University of Porto - Faculty of Engineering
Originally presented on: 11/6/2013 8:51:00 - 9:12:00

The PermSMBR is an innovative technology that consists in the integration of permselective membranes with the Simulated Moving Bed Reactor (SMBR)[1]. It was already evaluated for the production of oxygenated compounds, like acetals (1,1-diethoxyethane and 1,1-dibutoxyethane) and a green solvent (ethyl lactate). The PermSMBR shows a productivity enhancement accompanied by a significant reduction on the solvent consumption when compared with other intensified technologies, as reactive distillation and/or SMBR, proving PermSMBR high potential[2-3]. However, in those studies an integrated PermSMBR was considered, which comprises a set of tubular membranes packed with a solid that acts as catalyst and selective adsorbent to water (by-product). In this work, a new configuration is presented, the coupled PermSMBR, where the tubular membranes are located after fixed-bed columns packed with the catalyst/adsorbent. By this way the membranes will not be in contact with the solid, which from an industrial point of view will be easier to implement since the process of membranes installation/replacement and clean-up is simpler. The 1,1-dibutoxyethane production is used, as an example, and each set-up features are discussed. For the studied conditions, the highest productivity achieved using the coupled PermSMBR is 73.44 $Kg_{DBE}.L_{resin}^{-1}.day^{-1}$ with a desorbent consumption of 0.89 $L_{n\text{-}Butanol}/Kg_{DBE}$. The results obtained indicate the coupled PermSMBR suitability for the sustainable 1,1-dibutoxyethane production.

References:

(1) Silva, V. M. T. M.; Pereira, C. S. M.; Rodrigues, A. E. Simulated Moving Bed Membrane Reactor, new hybrid separation process and used thereof. PT Patent 104496; WO Patent 2010/116335, 2009.

(2) Silva, V. M. T. M.; Pereira, C. S. M.; Rodrigues, A. E. PermSMBR—A new hybrid technology: Application on green solvent and biofuel production. *AIChE Journal* **2010,** *57*, 1840-1851.

(3) Pereira, C. S. M.; Silva, V. M. T. M.; Rodrigues, A. E. Green Fuel Production Using the PermSMBR Technology. *Ind. Eng. Chem. Res.* **2011,** *51*, 8928–8938.

**Acknowledgement:** The research leading to these results has received funding from the European Union Seventh Framework Programme (FP7/2007-2013) under grant agreement n° 241718, EuroBioRef.

**(445c) 2-Acetoxy Propionic Acid Using Reactive Distillation â€“ An Intermediate in Synthesis of Bio-Acrylic Acid From Lactic Acid**

Kolah, A. K., Michigan State University
Lira, C. T., Michigan State University
Miller, D. J., Michigan State University
Originally presented on: 11/6/2013 9:12:00 - 9:33:00

Acrylic acid, its esters, and acrylate polymers are high volume chemicals and critical chemical building blocks used in a wide gamut of applications including superabsorbents, hard plastics, paints, detergents, and adhesives. The global acrylic acid market is estimated at 4.5 million tons and $12.4 billion annually in 2012, with a growth rate of about 4% per year.

Acrylic acid synthesized from biobased lactic acid is a drop-in replacement for the petroleum-based product synthesized from propylene. Lactic acid is a mature biobased feed stock with a proven track record of high yield (>95%) from glucose at a low cost.

Synthesis of biobased acrylic acid from lactic acid involves a two-step process. In the first step, lactic acid is reacted with acetic acid to form 2-acetoxy propionic acid (2-APA). In the second step of the process, 2-APA is cracked under high temperature to form acrylic acid. The first step reaction is thermodynamically unfavorable, with an equilibrium constant of 0.25, and results in poor conversions if carried out in a batch reactor even with high mole ratios of acetic acid to lactic acid.

In this work, the selective formation of 2-APA by reaction of lactic acid with acetic acid has been carried out using continuous reactive distillation. The reaction takes place in a six-meter tall, 50 mm ID Pyrex column operated at atmospheric pressure, and is catalyzed by Amberlyst-15 cation exchange resin supported inside Sulzer Katapak-SP11 structured packing. Lactic acid feed stocks from 20 wt% to 85 wt% in water were used in experiments along with glacial acetic acid. High conversion of lactic acid and its oligomers were observed (>98%). When 20% lactic acid was used, a selectivity of 96% to 2-APA was obtained.

An unusual observation in the experimental work was the enhancement of the acetic acid – water separation by the presence of lactic acid. Because of this enhancement, it is possible to obtain a distillate stream containing 99% water from the reactive distillation column while retaining acetic acid within the column where reaction occurs.Ternary VLE experiments were performed to obtain the NRTL parameters describing this system.

Experimental studies are complemented by process analysis using AspenPlus simulation software. By implementing a pseudo-homogeneous kinetic model generated in our laboratories from batch experimental kinetic data, and descsribing vapor-liquid equilibria based on experimental data, we are able to predict the experimentally observed pilot-scale column behavior.

It can be concluded that 2-APA can be synthesized very efficiently from aqueous lactic acid solutions by reaction with acetic acid using a continuous reactive distillation column.

**(445d) Novel Technology for Producing High Purity n-Hexane and Benzene**

Yu, Y., Tianjin University
Li, Y., Tianjin University
Geng, Z., Tianjin University
Ma, J., Tianjin University
Zhang, Z., Tianjin University
Zhang, M., Tianjin University
Originally presented on: 11/6/2013 9:33:00 - 9:54:00

N-hexane is one of the most representative non-polar solvent. It is widely used in pharmaceutical, chemical, polymer materials etc., and owning a great market demand. Benzene is one of the most important petrochemical materials. The yield and production technology of benzene stand for the petrochemical development level of the country. However, the traditional separation and purification technologies, for example, molecular sieve adsorption, batch azeotropic distillation and extractive distillation etc., are difficult to get high yield production of n-hexane and benzene with high purity. Moreover, the energy consumption is high in these processes. Our team has developed a series of new technologies coupling the normal distillation and extractive distillation to produce high purity n-hexane and benzene, and iso-hexane as a by-product. N-methyl pyrrolidone was chosen as the extractant in the process. C3~C8 alkane mixtures from raffinate of aromatics extraction unit, pyrolysis gasoline or straight-run gasoline were used as feed. Gradual separation was used to get the final products. Firstly, the iso-hexane was obtained after removal of the impurities with boiling point lower than 2-methylpentane. Then, the remaining materials were separated by extractive distillation to acquire n-hexane product. Finally, the benzene product was got by further separation. The multi-effect thermal coupling technique was employed in the distillation unit. With the heat integration and optimization, the usage of steam and cooling water was saved. The usage of extractant was also reduced by the way of the cycling between the extractive distillation column and the recovery column. To prevent impurities accumulated in the extractant, removal of the heavy component impurities was implemented by taking a small part of material into the falling film evaporation system. High purity n-hexane and benzene product were finally produced by the above mentioned method. The purity of n-hexane product is higher than 99 wt% with the content of benzene is lower than 100 ppm. At the same time, the concentration of benzene product is higher than 99.9 wt%. Currently, a 30,000 MTA n-hexane and benzene plant has been built up using this technology.

**(445e) Online Monitoring of Emulsion Polymerisations Using Raman Spectroscopy**

Houben, C., University of Cambridge
Nurumbetov, G., University of Warwick
Haddleton, D.,
Lapkin, A., University of Cambridge
Originally presented on: 11/6/2013 9:54:00 - 10:15:00

Many polymer products are manufactured using batch and semi-batch reactors. Typically the process parameters are fixed and are not varied during the reaction. Unexpected changes in operating conditions can lead to variations in polymer structure and purity, resulting in significant losses due to off-spec materials. An approach addressing the complex issues of real-time process control of large-scale emulsion polymerisation processes is to use advanced models and on-line sensors. A specific objective is to implement a model-based predictive control based on Raman spectroscopy.

## (445f) Process Intensification in Millistructured Extraction Columns

Holbach, A.,
Caliskan, E., TU Dortmund University
Kockmann, N., TU Dortmund University
Originally presented on: 11/6/2013 10:15:00 - 10:36:00

Process intensification in millistructured extraction columns

The downstream process of biochemical and pharmaceutical products can account up to 80% of the cost of production. [1,2] To reduce the cost it is necessary to develop new process intensified equipment with higher time-volume-yield. Process intensification is one important part to make the industry more economical and ecological. Some mass or heat transport limited chemical reactions are only technical possible or cost-efficient with process intensified equipment. The small channels and structures in this equipment gives the potential to control the fluids on the sub-millimeter scale. The known characteristics of transport phenomena enable an easy and fast scale up for bulk chemicals from lab to industrial scale.

We investigated the process intensification in millistructured extraction columns with an inner diameter of 15 mm, see figure 1. The column has 20 stirred cells and an active extraction high of 440 mm. The total internal volume with head and bottom is 220 ml with total height of 760 mm. With stirred cells we achieved high energy input, high specific surface between the two phases and thus good mass transfer. The backflow was decreased with narrow plates between the plates. Thus, the dispersed phase coalesces at the plates and plugs the stirred cells. With axial pulsation of the fluid the coalescence area is destroyed and the dispersed phase is transported into the next cell. This new technique allows for controlled flow of the dispersed phase into the next stage. The retention times for counter current processes can be adjusted for transport limited processes or reactive extraction systems.

The hydrodynamics and mass transfer inside the column was characterized with the European Federated Chemical Engineering (EFCE) testsystem n-butylacetate(d)/acetone/water(c). The mass transfer from continuous-to-disperse (c-d) and from dispers-to-continuous (d-c) and the droplet size distribution with extraction efficiency was investigated. Different to conventional columns, the energy input inside the stirred cells are higher and the influence on the coalescenc and mass transport direction is not so large. We achieved specific surfaces between the two fluids up to 3200 m^2/m^3 and measured the mass transfer up to 7 theoretical extraction stages per active high (H = 440 mm). This correspondsto 16 theoretical stages per meter, see figure 1. The experiments show that higher total throughput yields higher extraction efficiencies, hence, high mass flow rates are preferred.

The enantioselective extraction of (R/S)-α-cyclohexyl-mandelic acid will be investigated as an example; see the molecules in figure 2. The enantioselective extraction is an organic/aqueous reactive extraction. The enantiomers create in the aqueous phase a complex with hydroxy-propyl- -cyclodextrin. The (S)-complex is preferred to the (R)-complex, so that the extraction is selective. The whole process is mass transport limited and has a low selectivity. In conventional columns the extraction is impossible, because the retention time of the dispersed phase and the specific surface between the two fluids are too small. In process intensified extraction columns with high specific surfaces and many extraction stages the counter current extraction is possible. First results show a considerable separation of the two enantiomers.

[1] Lowe C. R., Lowe A. R., Gupta G., New developments in affinity chromatography with potential application in the production of biopharmaceuticals, J. Biochem. Biophys. Methods 49, 2001, 561?574

[2] Strube J., Sommerfeld S., Challenges in biotechnology production?generic processes and process optimization for monoclonal antibodies, Chemical Engineering and Processing, 44, (10), 2005, 1123-1137

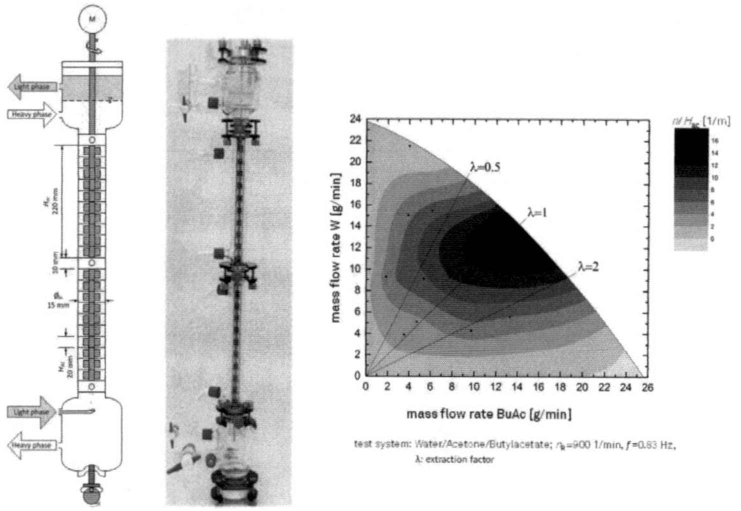


Figure 1: Millistructured extraction column; left side: sketch and picture of the column with two sections; right side: experimental extraction efficiency with different mass flow rates

(S)-α-cyclohexyl-madelic acid

(R)-α-cyclohexyl-madelic acid

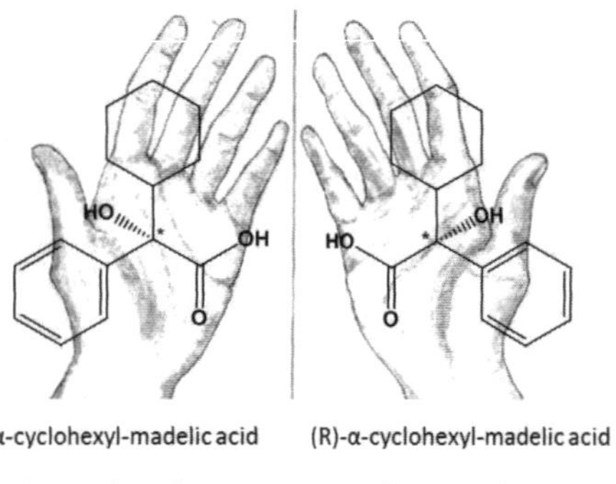


Figure 2: Molecular structure of (S) and (R)- α-cyclohexyl-madelic acid

**(445g) A NOVEL Comparison Among Intensified and Traditional Complete Ethyl Acetate Productions Designs**

Santaella, M., Universidad Nacional de Colombia
Narvaez, P. C., Universidad Nacional de Colombia
Orjuela, A., Universidad Nacional de Colombia
Originally presented on: 11/6/2013 10:36:00 - 10:57:00

Over the last decade different proposals of process intensification have been made for the improvement of the esterification reaction between acetic acid and ethanol. The dissimilar considerations and in some cases unrealistic assumptions made by the authors during modelling limit the possibility to compare among different production processes. In addition to this, different inlet compositions and different product specifications represent a major barrier to ensure an objective comparison between different processing technologies.

In this paper, the conventional production process, the reactive distillation (RD) and the reactive dividing wall column (RDWCD) process for the production of urethane grade ethyl acetate (99,5%) have been simulated using a commercial software (Aspen One 7.2). Glacial acetic acid and azeotropic ethanol were used as raw materials, and different catalysts were evaluated (self-catalytic, sulphuric acid, Amberlyst 15, Amberlyst 70). A validated NRTL-HOC thermodynamic model was used to describe phase equilibria. Chemical kinetic models were obtained from literature. A preliminary evaluation of reactive residue curve maps obtained with different catalysts and residence time ranges between 0-500min allowed to compare feasible column profiles.

Capital costs and energy savings for reactive distillation are predicted between 15-80% over classical configurations of a reactor followed by separation units. Dividing wall column, Petluk column and similar thermally coupled configurations predict 30-50% energy savings over conventional two distillation column processes. The RDWCD also predict 30-50% energy reduction over simple RD.

**(478a) Enhanced Yield of Bio-Gas Production With Modified Themo-Reactor Configuration and Separation of CO2**

Tahir, M. S., University of Gujrat
Saleem, M., Institute of Chemical Engineering and Technology, University of the Punjab, Lahore, Pakistan
Originally presented on: 11/6/2013 12:30:00 - 12:50:00

Pakistan is currently facing severe energy crisis with significant shortfall. A number of short-term technical solutions are available but are not adequate to solve the issue in the longer run. Production of bio-gas with modified thermo-reactor from the animal waste at regional level shows potential of reducing burden on total natural gas resource in Pakistan.

This research study investigates the effects of solar thermo-rector configuration on the overall yield and composition of bio-gas from dung. The designed capacity of the bio-gas plant is 10 $m^3$. The solar thermo-reactor employs internal heating system with coils using warm water heated from the solar energy. The bio-gas reactor was operated over the range of 25-55 $^0$C to observe effects on bio-gas yield and composition. It was observed that the increase in temperature within the thermo-reactor resulted in significant increase in yield of bio-gas i.e. thermo-reactor operated at 50-55 $^0$C (thermophilic). Operation time and plant size of state of the art process can be reduced.

The calorific value of bio-gas obtained at 50-55 $^0$C was slightly higher than calorific value of bio-gas obtained at mesophilic condition (25-35°C) due to increased production of $CH_4$. Thermophilic operation of bio-gas plant also enhances the methane content in the bio-gas i.e. 57-60 % vol of methane. Bio-gas produced was purged from bottom and cold water from top a packed column and decreases in $CO_2$ was also recorded. The results were scaled up from pilot scale capacity to regional level in order to estimate potential of bio-gas production

**(478b) Microwave-Assisted Catalytic Glycolysis of Poly(ethylene terephthalate) By Using Ionic Liquids**

Liu, Y., Donghua University
Xing, Y., Donghua University
Originally presented on: 11/6/2013 12:50:00 - 13:10:00

Poly(ethylene terephthalate) (PET) was widely used for beverage bottles, electrical/electronic instruments and household wares. As a consequence of dramatically increasing consumption, recycling of post-consumer PET products has become an important environmental opportunity for sustainable usage in society. The presented study is aiming to investigate the use of ionic liquids as catalysts for the glycolysis of PET. A series of novel ionic liquids using 1-butyl-3-methylimidazolium chloride (BMIMCl) and zinc halides (ZH) as precursor with different molar ratios was first synthesized. The glycolysis of PET by microwave-assisted method using the synthesized ionic liquids as catalysts and glycol as solvent was also studied. The influences of the type of zinc halide, the molar ratio of BMIMCl:ZH, the dosage of ionic liquid and solvent, reaction time and reaction time on the yield of recovered terephthalic acid were investigated. The synthesized ionic liquids and the glycolysis product were characterized by X-ray diffractometer, infrared spectrometric analyzer, $^{1}H$ NMR, differential scanning calorimeter and DSC. The results showed that the synthesized ionic liquids exhibited higher catalytic activity for the glycolysis of PET by microwave-assisted method□The halogen anion has a key influence on the glycolysis of PET. A study on the recycling of the ionic liquid shows that it could be used repeatedly. A mechanism of the degradation of PET was proposed and the kinetics of the reaction was needed further investigated.

**(478c) Degradation of Rhodamine-B By Heteropolyoxometalate Ionic Liquid Under Visible Light**

Liu, Y., Donghua University
Chen, Y., Donghua University
Xing, Y., Donghua University
Originally presented on: 11/6/2013 13:10:00 - 13:30:00

The photo-catalytic degradation of Rhodamine-B(RhB) dye was successfully achieved using different heteropolyoxometalate phosphonium ionic liquids(PWPILs) under visble light irradiation. A series of heteropolyoxometalate phosphonium ionic liquids were first prepared from phosphotungstic acid and different quaternary phosphonium salts trihexyl(tetradecyl)phosphonium bromide, tributyl(octyl)phosphonium bromide and tetrabutyl phosphonium bromide with different molar ratioa. The melting points of PWPILs, all lower than 100°C,were characterized by differential scanning calorimeter(DSC). The results show that the characterized peaks of phosphotungstic anion in IR spectra of PWPILs are consistent with that of phosphotungstic acid. This indicates that the quaternary phosphonium cations do not change the Keggin structure of heteropoly acid anion. The influences of experimental parameters, such as the carbon chain length in phosphoniums, the number of free hydrogen in phosphotungstic anions ,addition of different catalysts, irradiation time and the concentration of RhB solution on the photo-catalytic degradation of RhB were investigated. The photo-degradation efficiency was characterized by decolourization rate of RhB. The results indicated that $[P(C_{14}H_{29}(C_6H_{13})_3]_3[PW_{12}O_{40}]$ showed the best photo-catalytic degradation efficiency.
The RhB solution (2×10-4mol/L) was almost decomposed catalyzed by $[P(C_{14}H_{29})(C_6H_{13})_3]_3[PW_{12}O_{40}]$ under UV-Vis light irradiation for 3 hours. The degradation rate was more than 95%.

**(478d) Metal-Containing Ionic Liquid As An Effective Catalyst for the Degradation of Poly(ethylene terephthalate) in Ethylene Glycol**

Lu, X., Institute of Process Engineering, Chinese Academy of Sciences
Wang, Q., Institute of Process Engineering, Chinese Academy of Sciences
Zhang, S., Institute of Process Engineering, Chinese Academy of Sciences
Originally presented on: 11/6/2013 13:30:00 - 13:50:00

The vast emissions of poly(ethylene terephthalate) (PET) wastes have caused serious environmental pollution and resource waste. Its effectively recycling has become a global problem which must be urgently solved.

Chemical recycling is an attractive process because PET can be degraded into monomers which can be used as raw materials to produce virgin plastic products. Glycolysis, as an important chemical recycling method, has received extensive concern due to its advantages of less volatile solvents, high conversion of PET, giving monomers and easily continuous production. But there are some drawbacks, such as high temperature, high pressure, low reaction rate and difficult separation between catalyst and product. Many researchers devoted to developing new catalysts to overcome the drawbacks above. Ionic liquids have been reported that could effectively catalyze the degradation of PET to give BHET monomers with high purity and make the reaction condition mild. But the conversion of PET and the selectivity of monomer bis(2-hydroxyethyl terephthalate)(BHET) are both low, and the reaction rate is slow. Therefore, the development of new catalysts is still critical.

It has been found that synergetic effect of catalysts can speed up a chemical reaction or make reaction condition mild.[1,2] Yamamoto et al. found that combinations of organic and inorganic molecules bearing acidic groups could result in higher catalytic activity than individual catalysts for acid-catalyzed reactions.[2] Wang et al. and Zhou et al. found that the synergetic effect of cation and anion of ionic liquid catalyst could make the degradation condition of PET milder.[3,4] Therefore, we infer that a new catalyst containing two active components maybe has highly catalytic activity.

In this study, a metal-containing ionic liquid catalyst including two active components was synthesized and used for the PET degradation in ethylene glycol. The results showed that the synthesized catalyst has better low-temperature catalytic activity. The conversion of PET and the selectivity of BHET are 100 % and 78.7 % respectively at 175 Â°C for 1.25 h. The high catalytic activity is mainly attributed to the synergetic effect of two components. Moreover, this catalyst has the advantages of mild reaction condition, fast reaction rate, products being easily separated from catalyst and being of high purity.

[1] P. R. Schreiner. Science, 2010, 327, 965.

[2] H. Yamamoto and K. Futatsugi. Angew. Chem. Int. Ed., 2005, 44, 1924.

[3] H. Wang, R. Y. Yan, Z. X. Li, X. P. Zhang and S. J. Zhang. Catal. Commun., 2010, 11, 763.

[4] X. Y. Zhou, X. M. Lu, Q. Wang, M. L. Zhu and Z. X. Li. Pure Appl. Chem., 2012, 84, 789.

**(478e) Anaerobic Digestion Technology:The Energy,Environment and Health of Developing Nations**

Adetule, O., Cornell University

Originally presented on: 11/6/2013 13:50:00 - 14:10:00

Anaerobic digestion is one of the most cost effective and established technologies available for managing organic wastes for energy production,environmental management and pathogen reduction. In developing countries such as Nigeria, the application of this technology at household level as a method for gas generation and safe organic waste management has not been adequately studied. In light of the above, a pilot scale digester was designed for the production of Biogas which at the same time manages waste and reduced pathogens through anaerobic digestion process. The pilot scale digester was 10 litres and manages 9 litres of waste.The digester was fed with different types of waste:Cattle manure, Chicken litter, Rumen of cattle and Water hyacinth. After some weeks, the wastes produced Biogas, reduced pathogens and reduced the amount of wastes in the environment. A prototype digester was later designed and constructed which can accommodate more waste and produced more biogas. The prototype digester is a $5m^3$ underground fixed dome biodigester made of concrete and was designed, constructed and used for digesting cattle dung and water hyacinth. The digester was operated by mixing two parts of cattle dung (900kg) and one part of chopped water hyacinth (100kg) in 2,000 liters of water for two weeks following standard procedures. The gas produced was collected and piped to a kitchen for household use. It was also observed that there was reduction in pathogens and sanitization of the environment through waste management. This study shows that anaerobic digestion can be maximized for waste reduction, energy production,pathogen reduction and conversion of green house gases into usable forms.

**(478f) Phosphorus Recovery From Municipal Wastewater Treatment Sludge**

Lamichhane Upadhyaya, K., Mississippi State University
French, W. T., Mississippi State University
Hernandez, R., University of Louisiana at Lafayette
Originally presented on: 11/6/2013 14:10:00 - 14:30:00

## Phosphorus recovery from municipal wastewater treatment sludge

Kamal Lamichhane Upadhyaya[1], W. Todd French[1] and Rafael Hernandez[2]

[1]Dave C. Swalm School of Chemical Engineering, Mississippi State University, MS, 39762, USA

[2]Department of Chemical Engineering, University of Louisiana at Lafayette, Lafayette, LA, 70503, USA

### Abstract

Phosphorus is an essential nutrient for plants, animals and humans. Phosphorus is primarily used in agriculture, as it is one of the important components in fertilizer. Besides, phosphorus is also used in laundry detergent, matchstick, steel production and pesticides. The United States phosphate reserves are declining, which could lead to phosphorus shortage in near future. The United States now imports 10% of its phosphorus supply after exporting for several decades. Hence, there is need to search for renewable and sustainable source of phosphorus such as municipal wastewater sludge. Municipal wastewater sludge contains up to 10 wt% phosphorus and recovering these could significantly decrease the phosphorus shortage in the United States. In this study, the potential of phosphorus recovery from primary and secondary sludge through acidic and basic leaching processes will be investigated. The wastewater treatment sludge will be incinerated to obtain sewage sludge ash (SSA) and subjected to acidic and basic leaching at different acid and base concentration and pH levels. After leaching process, the supernatant will be separated from residue through filtration. Phosphorus will then be recovered from the supernatant mixture through the addition of calcium chloride in basic medium. The recovered phosphorus ($Ca_5(PO_4)_3OH$) has various agricultural and industrial usage as it resembles the phosphate mineral ore (apatite).

**(478g) Techno- Economic Analysis of Wastewater Biosolids Gasification**

Lumley, N. P. G., Colorado School of Mines
Porter, J. M., Colorado School of Mines
Originally presented on: 11/6/2013 14:30:00 - 14:50:00

Wastewater treatment in the U.S. produces over eight million dry tons of sludge per year. Wastewater sludge in its raw form is a noxious dilute suspension of organic matter in up to 95% water. Management and disposal is an expensive challenge for wastewater treatment plants (WWTPs), where the conventional approach to sludge management is to dewater the material to varying degrees and dispose by landfilling or application as a soil amendment. Such an approach requires energy and financial expense both in drying and ultimate disposal. Dried wastewater sludge has a lower heating value (LHV) close to that of low- rank coal. Thermochemical conversion of wastewater sludge to energy and/ or fuel product may allow for a more value- added approach to sludge management in addition to concentrating and stabilizing the waste stream.

Many WWTPs currently valorize a portion of the energy in sludge via anaerobic digestion. The biological process allows bacteria to operate on dilute sludge in an anaerobic environment producing a combustible bio- gas. Anaerobic digestion is an established technology but allows for only about 20% recovery of the energy in sludge. Anaerobic digestion does not significantly reduce the amount of sludge requiring downstream disposal. Additionally, bio -gas contaminants, especially siloxanes, are a problem for downstream combustion apparatus.

Thermochemical conversion (TCC) technologies subject sludge to chemical processes at elevated temperatures producing heat and/or combustible fuel products. TCCs applicable to wastewater treatment include direct combustion, pyrolysis, and gasification. TCC offers a great advantage over digestion by allowing nearly full utilization of sludge energy and a similarly great reduction in disposable waste volume. Each of these TCCs has been successfully applied to biomass. Wastewater sludge however is fundamentally different from most biomass sources. Raw wastewater sludge may be up to 95% water and 30% ash (dry basis). Balancing moisture removal with reactor operating conditions is the most important consideration in whether TCCs can be net energy- and net economically-positive. Ash concentration and composition present certain challenges to reactor design. Additionally, wastewater sludge organic composition can be variable leading to uncertainty in fuel output.

This work reviews the current state of practice of various TCCs with application to sewage sludge. We discuss technical and economic challenges restricting greater implementation of TCC within the wastewater process. A thermodynamic model of a sludge gasification- electric generator system is built using performance characteristics of a generic, commercially available small biomass gasification platform. We perform energy and techno- economic analyses on sludge gasification to determine lower- limit WWTP capacity and as- delivered sludge moisture content for both air- blown and steam- blown gasification processes.

**(501a) Low-Temperature and Low-Pressure Fuel Hydrodesulfurization By Solid Catalyst Coupling With Ionic Liquids**

Li, C., Institute of Process Engineering
Zhang, S., Institute of Process Engineering, Chinese Academy of Sciences
Originally presented on: 11/6/2013 12:30:00 - 12:52:00

A new hydrodesulfurization (HDS) process for the fuels utilizing a solid catalyst coupled with ionic liquids (ILs) at mild operating conditions is first proposed in this study. A novel Ce-Ni-Mo/γ-$Al_2O_3$ HDS catalyst with high catalytic activity was developed and coupled with different ILs to investigate the removal of dibenzothiophene (DBT) from model oil at low temperature and low pressure. Results showed that the sulfur removal efficiency follows the order: $[BDMIM]BF_4$> $[BMIM]BF_4$>$[EMIM]BF_4$> [BMIM]Cl. In addition, the highly efficient $[BMIM]BF_4$ coupled with the solid catalyst was selected to undergo a series of parallel experiments to systematically investigate the effects of temperature, hydrogen pressure and IL/oil mass ratio on DBT removal from model oil. To achieve optimal sulfur removal efficiency, model oil was conducted at the following conditions: reacton temperature, 150℃; hydrogen pressure, 3MPa; reaction time, 3h; rotate speed, 150rpm/min; and IL/oil mass ratio, 1:6. The process and mechanism of HDS were also studied using a molecular simulation software. This study shows that the HDS process using the solid catalyst coupled with ILs provides a new direction for producing cleaner industrial fuels.

**(501b) Reactivity of Biopolymers in Ionic Liquids**

Hallett, J. P., Imperial College
de Gregorio, G., Imperial College
Ismail, N. L.,
Originally presented on: 11/6/2013 12:52:00 - 13:14:00

Ionic liquids are gaining increasing acceptance as novel solvents for chemical transformations. This has included a recent focus on the use of ionic liquids for the separation of biopolymers (lignin and cellulose). During this process, the biopolymers are susceptible to multiple types of interactions with the ionic solvent, which can lead to depolymerisation (fragmentation or cleavage reactions), solvolysis or solvent addition reactions.

Our recently proposed Ionosolv process focusses on the use of highly acidic ionic liquids for biomass pretreatment. During pretreatment, all biopolymers in solution (lignin, cellulose, hemicelluloses) are subjected to harsh chemical consitions (non-aqueous, highly acidic, moderate temperature) which can induce novel reactivities. This paper will explore several potential reactions of the ionic liquid solvents with these biopolymers, including ether and ester cleavage and anion-induced substitution reactions. A variety of chemical characterisation techniques have been explored to ensure both a qualitative and quantitative measure of this reactivity.

We have also utilised a series of model compounds (esters, ethers and alcohols) to explore the reactivity of individual polymer subunits with the solvent under a variety of consitions. This portion focusses on electronic effects of the substrate and explores reaction kinetics and changes in mechanism with different functional groups.

**(501c) Microwave-Assisted Ionic Liquid Pretreatment and Catalysis of Lignin to Phenol in a Biphasic Reaction System**

Xu, J.,

Originally presented on: 11/6/2013 13:14:00 - 13:36:00

**Microwave-assisted Ionic Liquid Pretreatment and Catalysis of Lignin to Phenol in a Biphasic Reaction System**

**Yongshui Qu, Hao Luo, Jian Xu***

National Key Laboratory of Biochemical Engineering, Institute of Process Engineering, Chinese Academy of Sciences, Beijing 100190, People's Republic of China

*Corresponding author: jxu@home.ipe.ac.cn

**Abstract:** Recent concerns about the gradual depletion of conventional fossil resources and the pressure from global climate change have accentuated the need for new alternative resources. As one of the main components in biomass, lignin is the second most abundant natural polymer after cellulose which has the potential to serve as a sustainable source of energy and organic carbon to replace petroleum-based chemicals. Efficient conversion of lignin into high value-added chemicals is a key to improve the economic feasibility of biomass refinery. In the present study, a biphasic reaction system assisted by microwave heating is employed in the conversion of lignin into phenol, a vital petroleum-based building blocks used in production of fine chemicals and plastics. In order to understand the relationship among the structure of the ionic liquid, its effect on the change of lignin structure and yield of phenol, a series of functional ionic liquids are prepared to pretreat the industrial lignin and the changes on the structures and characteristics of the lignin (functional groups, microcosmic morphology, etc.) in the pretreatment and its effect on the phenol production are investigated. The study on the directional conversion of the pretreated lignin to phenols in a biphasic system is subsequently carried out by using the microwave irradiation instead of conventional heating, replacing the common catalysts (acids or alkalis) with the ionic liquids. The result shows that the phenol yield of 8.14% is achieved with the microwave irradiation at 400 W for 60 min and the produced phenol can be separated easily by a separation-friendly biphasic system. In this biphasic system, the industrial lignin pretreated by 1-ethyl-3-methylimimidazolium acetate ([EMIM]OAc) is catalyzed into phenol by an alkaline ionic liquid. The phenol product is continuously extracted into an organic phase consisting of methylisobutylketone (MIBK) and 2-butanol.

**Keywords:** Lignin; Ionic liquid; Pretreatment; Phenol; Catalysis

**(501d) Short- and Long-Term Thermal Stabilities Of Imidazolium-Type Ionic Liquids On Catalytic Metal-Oxide Supports**

Akcay, A., Koc University
Balci, V., Koc University
Uzun, A., Koc University
Originally presented on: 11/6/2013 13:36:00 - 13:58:00

Ionic liquids (ILs) have been attracting tremendous attention in the last few decades because of their unique physicochemical properties, such as negligible vapor pressure, high thermal stability, high solvating capacity, and wide ranges of density and viscosity. With these tunable characteristics, they have become the first choice in many applications such as organic synthesis, electrochemistry, separation, and especially in catalysis.

In catalysis field, two directions of IL application recently discussed in the literature are particularly attractive: *supported ionic liquid phase* (SILP) catalysts and *solid catalysts with ionic liquid layer*(SCILL). Here, the utilization of ILs in these catalysts not only provides opportunities to control and stabilize the nuclearity and morphology of the active sites, but also offers the following opportunities to improve the catalytic performance: i) ILs can impose a positive influence on the "chemical" properties of the catalysts by acting as a co-catalyst, either as an acidic function in a bifunctional catalyst or as a modifier; ii) ILs can control the effective concentration of reaction intermediates and stabilize charged transition states or ligands to control the reaction pathways. SILP type catalysts, where active species dissolved in IL, act as a homogeneous system immobilized on a high surface area porous solid support offering the benefits of heterogeneous systems. Since support enables a higher surface area and IL enables the better utilization of catalyst particles in the pores, the system results in the higher selectivity and activity compared to traditional catalysts. Similar to the SILP catalyst concept in SCILL applications, a supported metal catalyst is coated with IL layer.

Although SILP and SCILL concepts offer improved activity and selectivity, applications of these IL-assisted supported catalysts are limited with the thermal stabilities of ILs. Thermal stability limits determined for pure ILs are not applicable for these systems. Because, ILs directly interact with the metal-oxide supports and these interactions influence the thermal stability of ILs significantly. Therefore, elucidation of structural factors determining the thermal stability limits of ILs on metal-oxide supports is crucial for the selection of optimum IL for the application conditions of IL-assisted supported catalysts. Here we report the *short-term* and *long-term* thermal stability limits of 34 different imidazolium-based ILs coated on commonly used metal-oxide supports, $SiO_2$, $\gamma$-$Al_2O_3$, MgO. Structures of imidazolium ILs were varied systematically and the acidity of the support material increased from MgO to $\gamma$-$Al_2O_3$ and to $SiO_2$.

*Short-term* thermal decomposition temperatures of ILs supported on metal-oxides were determined by thermogravimetric analysis (TGA) at a linear heating rate of 10 K/min from 298 to 1073 K from the onsets of the derivative per cent weight changes. Results show that thermal stability of pure ILs significantly changes upon coating on metal-oxides with different point of zero charges (PZC) such as < 2.82 for $SiO_2$, 6-9.3 for $\gamma$-$Al_2O_3$ and 11.6-12 for MgO. Pure *1-ethyl-3-methylimidazolium tetrafluoroborate* [EMIM][$BF_4$], for example, is reported to be stable up to 723 K; however, our data illustrate that it starts to decompose at 609 ± 3, 534 ± 3 and 446 ± 3 K when coated on $SiO_2$, $\gamma$-$Al_2O_3$, and MgO, respectively. In general, imidazolium type ILs lose their thermal stability significantly as the acidity of the support decreases from $SiO_2$ to MgO. However, ILs with high acidity levels such as those with *dicyanamide* [DCA]$^-$ anion, *1,3-dimethoxyimidazolium bis(trifluoromethylsulfonyl)imide* [DMOIM][$NTf_2$] *and 1,3-dihydroxyimidazolium bis(trifluoromethylsulfonyl)imide* [DHIM][$NTf_2$] and *1-ethyl-3-methylimidazolium hexafluorophosphate* [EMIM][$PF_6$] do not follow this general trend and show higher thermal stabilities on MgO than on $SiO_2$. Similarly, effects of increase in the length of alkyl groups attached to the imidazolium ring depend strongly on the anion type too. For instance, thermal decomposition temperatures of [EMIM][$BF_4$], *1-butyl-3-methylimidazolium tetrafluoroborate* [BMIM][$BF_4$], *1-hexyl-3-methylimidazolium tetrafluoroborate* [HMIM][$BF_4$] and *1-decyl-3-methylimidazolium tetrafluoroborate* [$C_{10}$MIM][$BF_4$] were determined on the $SiO_2$, $\gamma$-$Al_2O_3$, MgO metal-oxide supports. Results show that thermal stabilities of these ILs decrease from 609 ± 3 to 502 ± 3 K on $SiO_2$, from 534 ± 3 to 496 ± 3 K on $\gamma$-$Al_2O_3$ and from 446 ± 3 to 440 ± 3 K upon step-wise increase in the length of alkyl chain attached to the imidazolium ring from *ethyl* to *decyl*. However, when [$BF_4$]$^-$ anion is replaced with [$NTf_2$]$^-$, increase in the alkyl chain length of cation does not affect the thermal stability significantly. Similarly when the *butyl* chain was replaced with a nucleophilic *cyanopropyl* group on imidazolium ring with [$NTf_2$]$^-$anion, data does not indicate any effect on the thermal decomposition temperature on the same support.

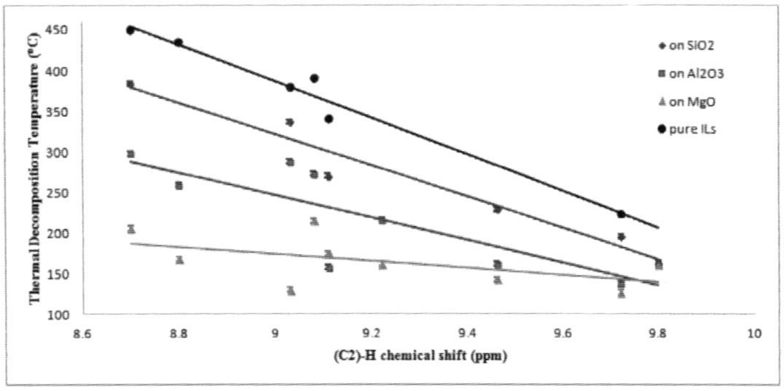


Figure 1. Thermal decomposition temperatures of several ILs with [BMIM]$^+$ cation as a function of (C2)-H chemical shift

As shown in Figure 1, thermal stabilities of ILs with [BMIM]$^+$ cation on various metal-oxides vary linearly with the chemical shift value of acidic (C2)-H on imidazolium ring. Moreover, slopes of these lines depend strongly on the PZCs of the metal-oxide supports. Thus, we infer that there is a strong correlation between the thermal stabilities of ILs with their acidity levels and the PZCs of supports.

In order to complement the TGA-based *short-term* thermal stability results, *long-term* thermal stability limits of ILs on metal-oxides were also investigated. For this purpose, ILs on various metal oxides were characterized by FTIR spectroscopy and SEM analysis with EDX mapping before and after 6 hour-long isothermal treatments at various temperatures. Results illustrated that some ILs start to decompose earlier than the temperatures determined by quick TGA scans while others remain intact up to the TGA-determined decomposition temperatures. As an example, on $\gamma$-$Al_2O_3$ metal-oxide support, [EMIM][$BF_4$] is stable up to the 534 ± 3 K (temperature determined with TGA scans) while [$C_{10}$MIM][$BF_4$] starts to decompose after 473 ± 3 K, earlier than the TGA-determined temperature.

In summary, results presented here provide the basis for selecting stable ILs for IL-assisted supported-catalysts based on their application conditions.

**(501e) Cost-Efficient Ionic Liquid and Water Mixtures for Biomass Dissolution**

George, A., Joint BioEnergy Institute
Tran, K.,
Sathitsuksanoh, N., Joint BioEnergy Institute
Sun, N., Joint BioEnergy Institute
Mac Dowell, N., Imperial College London
Singh, S., Joint BioEnergy Institute, Emeryville, CA 94608 and Sandia National laboratories, Livermore, CA
Simmons, B., Joint BioEnergy Institute
Welton, T., Imperial College
Hallett, J. P., Imperial College
Originally presented on: 11/6/2013 13:58:00 - 14:20:00

The adoption of lignocellulosic feedstocks in an industrial process used to commercially produce biofuels is contingent upon cost-effectively and efficiently disrupting the lignin present, as it inhibits downstream enzymatic-hydrolysis efficiency. There are numerous reports in the scientific literature on the capability of ionic liquids (ILs) to disrupt cellulose structure and dissolve lignin effectively. Ionic liquids offer exciting prospects due to properties such as their low volatility and "tunability". In this study a new class of ionic liquids were designed and tested for their capacity to improve the enzymatic hydrolysis efficiency of switchgrass. The ionic liquids tested were based upon a sulphate anion, whilst varying the cation. The number of alkyl chains on ammonium based cations were investigated. The ionic liquids were designed so that they can effectivly remove lignin and leave the cellulose intact, thereby enabling easier fractionation post pretreatment. Another benefit of this class of ionic liquids is their lack of sensitivity to $H_2O$, enhancing their handling properties. A focus was also placed on designing these ionic liquids from ubiquitous low cost chemicals, with the aim of mitigating reagent cost, thereby enhancing their industrial relevance. The metrics used to evaluate these ionic liquids were degree of lignin dissolution, effect on cellulose structure, ease of pretreated material fractionation and enzymatic hydrolysis efficiency.

## (501f) Conversion of Residual Ethylene Oxide in the Ionic Liquid Catalyzed CO2 Cycloaddition Process

Sun, J., Institute of Process Engineering, Chinese Academy of Sciences
Cheng, W., Institute of Process Engineering, Chinese Academy of Sciences
Wang, J., Institute of Process Engineering, Chinese Academy of Sciences
Zhang, J., Institute of Process Engineering, Chinese Academy of Sciences
Zhang, S., State Key Laboratory of Multiphase Complex Systems, Institute of Process Engineering, Chinese Academy of Sciences
Originally presented on: 11/6/2013 14:20:00 - 14:42:00

Conversion of residual ethylene oxide in the ionic liquid catalyzed $CO_2$ cycloaddition process

Jian Sun, Weiguo Cheng, Jinquan Wang, Junping Zhang and Suojiang Zhang

*Institute of Process Engineering, Chinese Academy of Sciences,100190, Beijing, PR China*

**Contact information: jsun@home.ipe.ac.cn*

Recently, the synthesis of cyclic carbonates via cycloaddition of $CO_2$ to epoxides is much attractive not only because of its efficiency in $CO_2$ conversion, but also becuase of its widely derivative high-value chemicals. [1] Although a number of efficient catalysts including ionic liquids (ILs) have been developed, [2] attention on the trace residual epoxide in the reaction system, especially in gas phase, is rarely covered, which could cause a potential explosion of epoxide when unreacted $CO_2$ is compressed for recyle. As a continous work for $CO_2$ conversion,[3] herein, studies on the conversion of residual trace epoxide (0.2 wt%) in an IL catalyzed $CO_2$ cycloaddition process to produce ethylene carbonate (EC) were carried out. The effect parameters including pressure, temperature, time and solvents (e.g. $H_2O$, EC+ethylene glycol+$H_2O$, BmimCl, $BmimPF_6$, $BmimBF_4$, etc) were systematically screened, and a >99.9% conversion of EO could be realized by reduding the partial pressure of EO in the system. The results indicated that a vapor-liquid equilibriuma of EO in the reaction system has a more important effect on EO elimination rather than reaction time and temperature.

### Refferences

[1] B. Schaffner, F. Schaffner, S.P. Verevkin and A. Borner, ***Chem. Rev***., 2010, 110, 4554.

[2] a) J. M. Zhang, J. Sun, X. C. Zhang, Y. S. Zhao, S. J. Zhang. ***Greenhouse Gas Sci Technol***., 2011, 2:142-159; b) M. North, R. Pasquale, C. Young, ***Green Chem***., 2010, 12, 1514.

[3] a) J.Q. Wang, J. Sun, W.G Cheng, S.J. Zhang. ***Phys. Chem. Chem. Phys.***, 2012, 14, 11021; b) J. Sun, L. J. Han, W. G. Cheng , J. Q. Wang, X. P. Zhang, S. J. Zhang. ***ChemSusChem***, 2011, 4: 502; c) J. Sun, J.Q. Wang, W. G. Cheng, J. X. Zhang, X. H. Li, S. J. Zhang, Y. B. She. ***Green Chem***., 2012, 14: 654; d) J. Sun, W. G. Cheng, W. Fan, Y. H. Wang, Z. Y. Meng, S. J. Zhang. ***Catal. Today***, 2009, 148: 361.

### Acknowledgments

This work was supported by National Basic Research Program of China (2009CB219901), and National Science Fund of China (21006117).

**(501g) Aggregation and Solubilization Phenomena During Ionic Liquid Co-Solvent Based Lipid and Phorbol Ester Extraction From Biomass**

Bandlamudi, S., South Dakota School of Mines & Technology
Cooney, M., University of Hawaii
Benjamin, K. M., South Dakota School of Mines and Technology
Originally presented on: 11/6/2013 14:42:00 - 15:04:00

The use of ionic liquids to separate oils, proteins, and phorbol esters from biomass is a significant achievement in efficiently separating high quality jet and diesel fuels and value-added products from abundant ocean algae and other agricultural feedstock. Micro-algae and oil-seed biomass were suspended in co-solvents mixtures comprised of ionic liquids of varying polarity and polar-covalent molecules and subjected to mild heating under reflux at atmospheric pressure. Yields of bio-oils and phorbol esters extracted were determined by gravimetric analysis and HPLC, respectively. In addition, light scattering and fluorescence techniques were utilized to investigate polarization/anisotropy in the co-solvent mixture, and their effects on lipid solubility and rates of diffusion.

Molecular dynamics simulations have been employed to probe the fundamental interactions between solutes (lipids and phorbol esters) and ionic liquid and polar covalent molecules in solution. Simulations are capable of predicting the mixture density as a function of composition, and compare well with experimental data. Radial distribution functions (rdfs), potentials of mean force (pmfs), and simulation snapshots confirm the increased aggregation between IL cations and anions with increasing methanol concentration.

The aggregation behavior of triolein in IL/co-solvent mixtures was also explored. From the MD generated rdfs and potentials of mean force, we observe a clear correlation between the extent of triolein aggregation and the experimental extraction yields as a function of methanol composition. This indicates that the lipids are aggregating in these IL/co-solvent mixtures, a simulation finding that is also supported by light scattering experiments. Therefore, there is a clear correlation between the free energy of lipid aggregation and the overall extraction yield. Therefore, the dominant mechanism for the lipid auto-partitioning behavior is lipid aggregation, likely leading to the formation of micelles, which subsequently rise through the IL/co-solvent mixture via buoyancy (density difference).

In addition, density functional theory (DFT) calculations using the COSMO-RS implicit solvent model and the COSMO-THERM framework are used to predict phorbol ester solubilities in IL/methanol mixtures, which are subsequently compared to experimental ester extraction results.

Both experimental and simulation efforts provide an improved understanding of the relationship between molecular-level co-solvent structure and solute aggregation/solubilization, which will help in the design of tailored ionic liquid/PCM solvent mixtures to optimize lipid and ester extraction and separation.

**(519b) Biorefinery: Liquefaction of Pyrolysis Char to Biofuel**

Feiner, R., Graz University of Technology
Schwaiger, N., Graz University of Technology
Pucher, H., Graz University of Technology
Pucher, P., BDI BioEnergy International AG
Siebenhofer, M., Graz University of Technology
Originally presented on: 11/6/2013 12:52:00 - 13:14:00

Lignocellulosic biomass is the most abundant renewable material [1] in the world with coverage of 95% of all plant biomass [2] which makes it a feasible feedstock for chemical conversion to biofuels. For successful substitution of fossil fuel there are still obstacles to overcome. State of the art technologies of the petrochemical and chemical industry are not directly applicable for converting biomass to biofuels. This is due to the fact that biomass is a highly oxygenated raw material. Behrendt et al. specify the chemical composition of dry wood as $CH_{1.4}O_{0.7}$ [3] figuring out nearly one atom of oxygen per carbon atom, underlining the indispensable need of hydrogen for converting lignocellulosic biomass to biofuels. By these means simple and cheap pre-treatment is needed to eliminate oxygen in the feedstock. Liquid-phase pyrolysis is a feasible route for thermo-chemical elimination of oxygen from biomass. Products of liquid-phase pyrolysis are (1) *non-condensable gases*, (2) *water*, (3) *pyrolysis oil* of poor combustion properties and (4) biochar with an amount of 40wt% on dry biomass feed basis. The formation of liquid-phase pyrolysis products was investigated and reported [4], [5].

Biochar is a major product of pyrolysis processes, such as torrefaction and hydrothermal carbonization [6], [7]. Biochar from liquid-phase pyrolysis mainly consists of biogenous carbon with an approximate amount of 99 wt% [4]. This makes biochar a considerable feedstock for direct liquefaction in order to produce biofuels. In principle the direct liquefaction of biochar may well base on the experience from fossil coal liquefaction with hydrogen and/or hydrogen donor solvents at elevated temperature and pressure. The products of direct coal liquefaction are highly aromatic and suitable for usage as high-octane gasoline or feedstock for aromatic chemicals [8].

During coal liquefaction, while thermal decomposition ruptures the coal bonds, hydrogen from the gaseous phase or from a hydrogen donor solvent stabilizes the fragments and eliminates heteroatoms like oxygen. Stabilization ensures the breakdown of macromolecules into smaller hydrocarbon fractions. At low hydrogen availability fragments are not stabilized and will undergo re-polymerization reactions to form tar [9].

Liquefaction of biochar was investigated in a stirred 450ml batch reactor between 370°C and 425°C. The feed product was obtained from pilot scale liquid-phase pyrolysis from BDI (OMV refinery, Vienna). Biochar was extracted with n-hexane, dried and milled to <200microns. For each experiment 30g of dried biochar batches, mixed with tetralin, were prepared. The mixture was fed to the reactor and pressurized with 50bar of hydrogen. Temperature was elevated from ambient to final operation temperature with a heating rate of 15 °C/min. In order to distinguish between the contribution of different process phases to conversion and product formation the *heating period* was investigated separately prior to investigation of the *isothermal phase*. Reaction time for the *isothermal stage* was limited to 30min. Experiments were conducted with continuous hydrogen supply at two hydrogen pressure levels. Upon completion of the experiment the heating jacket was removed and the reactor was cooled with compressed air. The gas amount as well as the gas composition were determined when reactor pressure was relieved.

The lumped parameters gas (G), oil (O), asphaltenes (A), pre-asphaltenes (PA) and residue (R) were chosen for evaluation of the progress of liquefaction. The parameters were calculated from the results of sequential extraction of filtered liquefaction residues after each experiment. The liquid product phase was subjected to GC-MS, GC-FID, and elemental Analysis as well as size exclusion chromatography (SEC) in order to generate data for product quality, product formation, donor-solvent consumption as well as product size distribution.

Experiments with hydrogen donor at 425°C resulted in biochar conversion (C) of 84% and an oil yield of 72% after 30min of isothermal reaction, with the initial heating phase contributing to biochar conversion (C) to an extent of 37%. Biochar is converted into oil (O), gas (G), pre-asphaltenes (PA) and asphaltenes (A). Pre-asphaltenes and asphaltenes are still high molecular intermediates, formed during conversion of biochar [8], [10]. Liquefaction of biochar at greater extent starts at 415°C.

[1] X. Zhao, K. Cheng, and D. Liu, "Organosolv pretreatment of lignocellulosic biomass for enzymatic hydrolysis.," *Applied microbiology and biotechnology*, vol. 82, no. 5, pp. 815–27, Apr. 2009.

[2] R. Rinaldi and F. Schüth, "Design of solid catalysts for the conversion of biomass," *Energy & Environmental Science*, vol. 2, no. 6, p. 610, 2009.

[3] F. Behrendt, Y. Neubauer, M. Oevermann, B. Wilmes, and N. Zobel, "Direct Liquefaction of Biomass," *Chemical Engineering & Technology*, vol. 31, no. 5, pp. 667–677, May 2008.

[4] N. Schwaiger, R. Feiner, K. Zahel, A. Pieber, V. Witek, P. Pucher, E. Ahn, P. Wilhelm, B. Chernev, H. Schröttner, and M. Siebenhofer, "Liquid and Solid Products from Liquid-Phase Pyrolysis of Softwood," *BioEnergy Research*, vol. 4, no. 4, pp. 294–302, Jun. 2011.

[5] N. Schwaiger, V. Witek, R. Feiner, H. Pucher, K. Zahel, A. Pieber, P. Pucher, E. Ahn, B. Chernev, H. Schroettner, P. Wilhelm, and M. Siebenhofer, "Formation of liquid and solid products from liquid phase pyrolysis.," *Bioresource technology*, vol. 124, pp. 90–94, Nov. 2012.

[6] D. Ciolkosz and R. Wallace, "A review of torrefaction for bioenergy feedstock production," *Biofuels, Bioproducts and Biorefining*, vol. 5, pp. 317–329, 2011.

[7] A. Funke and F. Ziegler, "Hydrothermal carbonization of biomass: a summary and discussion of chemical mechanisms for process engineering," *Biofuels, bioproducts and biorefining*, vol. 4, no. 2, pp. 160–177, 2010.

[8] T. Kaneko, F. Derbyshire, E. Makino, D. Gray, and M. Tamura, "Coal Liquefaction," in *Ullmann's Encyclopedia of Industrial Chemistry*, Wiley-VCH Verlag GmbH, 2005, pp. 1–83.

[9] M. W. Haenel, "Catalysis in Direct Coal Liquefaction," in *Handbook of Heterogeneous Catalysis*, Wiley-VCH Verlag GmbH, 2008, pp. 3023–3036.

[10] X. Li, H. Hu, S. Zhu, S. Hu, B. Wu, and M. Meng, "Kinetics of coal liquefaction during heating-up and isothermal stages," *Fuel*, vol. 87, pp. 508–513, Apr. 2008.

**(519c) Mechanistic Insight Into Sonochemical Biodiesel Synthesis Using Heterogeneous Base Catalyst**

Choudhury, H. A., Indian Institute of Technology Guwahati
Moholkar, V., Indian Institute of Technology, Guwahati
Originally presented on: 11/6/2013 13:14:00 - 13:36:00

**Abstract**

The beneficial effect of ultrasound on transesterification reaction is well known. Heterogeneous (or solid) catalysts for biodiesel synthesis have merit that they do not contaminate the byproduct of glycerol. In this paper, we have attempted to identify the mechanistic features of ultrasound–enhanced biodiesel synthesis with the base–catalyst of CaO. A statistical design of experiments (Box–Behnken) was used to identify the influence of temperature, alcohol to oil molar ratio and catalyst loading on transesterification yield. The optimum values of these parameters for the highest yield were identified through Response Surface Method (with a quadratic model) and ANOVA. These values are: temperature = 62°C, molar ratio = 10:1 and catalyst loading = 6 wt%. The activation energy was determined as 82.3 kJ/mol, which is higher than that for homogeneous catalyzed system (for both acidic and basic catalyst). The experimental results have been analyzed vis–à–vis simulations of cavitation bubble dynamics. Due to 3–phase heterogeneity of the system, the yield was dominated by intrinsic kinetics, and the optimum temperature for the highest yield was close to boiling point of methanol. At this temperature, the influence of cavitation bubbles (in terms of both sonochemical and sonophysical effect) is negligible, and ultrasonic micro–streaming provided necessary convection in the system. The influence of all parameters on the reaction system was found to be strongly inter–dependent.

**(519d) Energetic and Environmental Assessment of Biomass Processing in the Sugar-Cane Industry**

Kahwaji Janho, M., Cleveland State University
Shuster, Q. A., Cleveland State University
Gatica, J. E., Cleveland State University
HernÃ¡ndez, M. R., Universidad Nacional de Tucuman
Mele, F. D., Universidad Nacional de Tucuman
Colombo, M., Universidad Nacional de Tucuman
Originally presented on: 11/6/2013 13:36:00 - 13:58:00

Bio-fuels have come to be recognized as possible, environmentally friendly alternatives to fossil fuels. This work studies a Bio-ethanol production process complementing a sugar mill, through computer simulations, using ASPEN PLUS® as the simulation engine.

The process flowsheet examined includes three major processing stages typical of distillery installations for fuel-grade production in typical sugar-cane processing facilities: (1) the fermentation stage, (2) the separation stage, and (3) the dehydration stage.

Using plant data from a bio-ethanol pilot plant in Northwestern Argentina (NOA) a kinetic model is formulated for a fed-batch molasses fermentation process. The model for this unit, along with a user-defined separation module, is then integrated into a commercial process simulator. Different schemes meeting equivalent performance standards in terms of purity and recovery are examined in this paper.

Particular attention is given to energetic integration by performing a pinch-analysis to each of the fundamental stages to find a global optimum. Optimization is achieved by minimizing a given objective function that accounts for bio-ethanol recovery, separation efficiency, and minimization of effluents. Life Cycle Analysis is used to complete the assessment of operating parameters/conditions on environmental effects.

**(519e) Investigation of Solvent System for the Production of Biodiesel From Sludge Palm Oil (SPO) By Enzymatic Transesterification**

Alam, M. Z., International Islamic University Malaysia
Nasaruddin, R. R., International Islamic University Malaysia
Jami, M. S., International Islamic University Malaysia
Uddin, M. S., International Islamic University Malaysia
Originally presented on: 11/6/2013 13:58:00 - 14:20:00

Biodiesel is a non-toxic, renewable and environmental friendly fuel that can replace the fossil fuels which have issues on toxic emissions, unstable escalation of price and high depletion in reserve. The high cost of common feedstocks of biodiesel can be reduced by using low cost and abundant waste oils such as sludge palm oil (SPO) especially in Malaysia which produces 40 million tons SPO per year. In this study, SPO with a very high FFA content (51.64 ± 0.59% as palmitic acid) was used to produce biodiesel using locally-produced *Candida cylindracea* lipase from a low-cost based medium, palm oil mill effluent (POME). The inhibition of lipase activity is reduced by studying the effects of several factors to the conversion of free fatty acids (FFA) into biodiesel and the total yield of biodiesel ($Y_T$).

In this study, sludge palm oil (SPO), a low grade and by-product of palm oil mill industry was used to produce biodiesel by direct enzymatic transesterification using locally produced *Candida cylindracea* lipases from palm oil mill effluent (POME). The solvent system was investigated by comparing the effect of methanol and ethanol. The yield of biodiesel and conversion of free fatty acids (FFAs) into biodiesel for ethanol were higher than the methanol. Ethanol was selected for the based on the findings of ethanol-to-SPO molar ratio, water content, and additional organic solvent. The optimum factors of ethanol solvent system were 4:1 ethanol/SPO molar ratio in the presence of small water content (1:15-1:10 water/SPO volume ratio). The optimum factors of process conditions were 10 U, $40^0$C, 250 rpm and 24 hours for enzyme loading, reaction temperature, mixing speed and reaction time respectively. The highest conversion and $Y_T$ of the optimal conditions from OFAT method was 12.0% and 62.3 % (w/w SPO) respectively.

**(519f) Update to Biodiesel Production From Activated Sludge and Economic Analysis**

Coker, A. T., Mississippi State University
Hernandez, R., University of Louisiana at Lafayette
Revellame, E. D., Mississippi State University
Holmes, W. E., University of Louisiana at Lafayette
French, W. T., Mississippi State University
Originally presented on: 11/6/2013 14:20:00 - 14:42:00

Biodiesel is a renewable fuel that can supplement petroleum fuel supply. One of the reasons why biodiesel is not yet cost-competitive with petroleum fuels is due to the high cost of feedstock - up to 70 -95% of biodiesel production cost. A feedstock with high potential for less expensive biodiesel production is activated sludge obtained from wastewater treatment plants. However, drying of sludge prior to oil extraction and reaction is a major operating cost of this process since it is so high in water content. Previous studies by this group have shown that the lipids present in sludge can be converted to biodiesel. In this case, supercritical methanol was used as a reactant and catalyst for biodiesel production.

A study was conducted to determine the optimum and cost-efficient yield of biodiesel produced from reacting wet sludge using only supercritical methanol. The goal of this investigation is to address the challenge of reducing sludge moisture content by investigating the feasibility of using activated sludge for cost-efficient biodiesel production with little or no drying of the feedstock. A model system of oleaginous yeast, *Rhodotorula glutinis*, was used to evaluate the production of biodiesel in a system similar to sludge. The combination of a fluidized sand bath and 46ml tubular stainless steel reactors were used for the supercritical methanol reactions. Two methods of this non-catalytic process, 1-step and 2-step processes, that could produce high FAME yields were studied and compared in terms of FAME yields and kinetic rate constants. With these results, plant schematics were proposed and economic analyses were conducted to investigate the cost efficiency of both non-catalytic methods. One was recommended with great potential for producing biodiesel from activated sludge at a cost-efficient price.

**(519g) An Intensified Integrated Biodiesel Manufacturing Process for Conversion of Waste Oils and Alcohols Into Biodiesel to Power Campus Bus System**

Sunol, A. K., University of South Florida
Cerniga, Z., University of South Florida
Mcgranaghan, E., University of South Florida
Originally presented on: 11/6/2013 14:42:00 - 15:04:00

A novel biodiesel manufacturing process is described. The continuous process manufactures biodiesel from waste oils and waste alcohols from local community and aims to power the University bus system. The process utilizes supercritical alcohols in transesterification reactions with residence times of few minutes. The autothermal reactor is heat integrated with the waste alcohol distillation process. The presentation reports the pilot process operation, process scale up, transporatation system powering implications and resulting life cycle analysis.

**(534a) Plate Food Waste: A Valuable Source of Energy**

Jung, S., Iowa State University

Originally presented on: 11/6/2013 15:15:00 - 15:35:00

Post-consumer food waste, because of its quantity and composition, could be a good feedstock for biofuel production. The effect of glucan loading (2 and 6%), addition of an enzyme cocktail, pretreatment (grinding/drying or sterilizing), and food waste composition (breakfast vs. lunch, presence of used napkins) on ethanol concentration was reported. The potential for sorting and converting used napkins into ethanol was also assessed. At a 2% glucan content, ethanol concentration of up to 16 g/l was obtained after saccharification/fermentation of breakfast and lunch food waste. Increasing glucan content to 6% significantly increased ethanol concentration to 54 g/L, while there was no significant effect of food waste composition on carbohydrate conversion efficiency. When converting ethanol concentration from g/L to L/g of waste, concentrations of 0.06, 0.06 and 0.40 were obtained for food waste, food waste w/o napkins, and used napkins, respectively, identifying used napkins as an excellent source of bioethanol production.

**(534b) Biofuel Production and Kinetics Study of Catalytic Microwave Pyrolysis of Douglas Fir Pellet Over Activated Carbon Supported Metal Catalyst**

Bu, Q., Washington State University
Lei, H., Washington State University
Wang, L., Washington State University
Tang, J.,
Originally presented on: 11/6/2013 15:35:00 - 15:55:00

The presented study is aiming to improve the quality of bio-oils by catalytic upgrading of pyrolysis oil from microwave pyrolysis of biomass using activated carbon (AC) supported metal catalyst. A central composite experimental design (CCD) was used to optimize the reaction condition of this process. The effects of reaction temperature and reaction time on product yields were investigated. GC/MS analysis indicated that the main chemical compounds of bio-oils were furans, phenols, guaiacols, and ketones/ethers which were in the range of 80-87% of bio-oils for most of the samples. The ketones/ethers mainly composed by 2-cyclopen-1-one and straight chain ethers were increased up to about 38% of bio-oil, and the phenols increased significantly accompanying with the decrease of guaiacols in comparison with raw biomass pyrolysis. Thermogrametric analysis (TGA) was used to analyze the thermal behavior or Douglas fir pyrolysis, and the kinetic parameters were investigated using thermogrametric analysis (TGA) data and compared with catalytic microwave pyrolysis of Douglas fir in this process, the activation energy of Douglas fir pyrolysis in this process was remarkable lower than conventional TGA pyrolysis of Douglas fir. And the reaction mechanism of this process was analyzed.

**(534c) Molecular-Level Modeling of Municipal Solid Waste Gasification**

Horton, S. R., University of Delaware
Zhang, Y., Air Products and Chemicals, Inc.
Bennett, C. A., University of Delaware
Klein, M. T., University of Delaware
Petrocelli, F., Air Products and Chemicals, Inc
Originally presented on: 11/6/2013 15:55:00 - 16:15:00

**Molecular-level Modeling of Municipal Solid Waste Gasification**

**Scott R. Horton*, Craig A. Bennett, Michael T. Klein**

*Department of Chemical & Biomolecular Engineering, University of Delaware*

**Yu Zhang, Francis P. Petrocelli**

*Air Products and Chemicals, Inc.*

In 2010, there was 250 million tons of municipal solid waste (MSW) in the United States. Over half of this waste ended up in landfills which fail to recover the energy stored within MSW. The most basic waste-to-energy technology is incineration. While this process recovers energy, it also poses problems to the environment due to $SO_x$, $NO_x$, dioxin, and furan emissions. Plasma-arc gasification has nonstoichiometric amounts of oxygen and reduces the production of these harmful oxygenates while recovering energy from waste via syngas production. Furthermore, due to the extremely high temperatures of the plasma torch, the final by-product is a salable glass-like slag. The objective of this study is to aid process design by developing a molecular-level kinetics model of plasma-arc gasification.

The MSW is typically described at the lumped level, i.e., biomass, food waste, and plastics. In the present work, each of these lumped components was described at the molecular-level. Biomass was modeled as cellulose, hemicellulose, and lignin. In addition to these molecules, lipids, common amino acids, and sugars were used to model food waste. The modeled plastics were polyethylene, polyethylene terephthalate, and polyvinyl chloride. These feeds were all represented at the molecular level using Flory statistics for nearly linear polymers and lattice theory for lignin.

The feed stream was converted into a reaction network using an in-house software tool, the Interactive Network Generator (INGen). There were two major categories of reactions: devolatilization/pyrolysis and gasification. The major pyrolysis reaction families were thermal cracking, decarboxylation, and decarbonylation. For gasification, the dominant reaction type was reaction with oxygen. The generated reaction network was solved and optimized using another in-house software, the Kinetic Modeling Editor (KME).

For integration with process simulation, the optimized KME model was integrated into ASPEN PLUS using user-defined models. In the case of gasification, the KME model contains reactors in series and rate laws employing linear-free energy relationships for reaction families. The process of KME-Aspen integration was generalized allowing for the automatic implementation of any reaction model into an ASPEN worksheet.

The KME model could also be coupled to a computational fluid dynamics (CFD) solver and take equipment level information into account in the design. Once the products and kinetics of the pyrolysis and gasification reactions are obtained from KME, the reaction network could be converted into one or several global pyrolysis/gasification reactions by kinetic model reduction. The simplified mechanism could be easily integrated into any CFD solver, such as Ansys FLUENT. On the other hand, equipment information obtained from the CFD simulation, such as flow and temperature pattern, provides guidance for choosing reactors in ASPEN PLUS process simulation.

**(534d) Performance Analysis of Rdf Gasification in a Two Stage Fluid Bed - Plasma Process**

Materazzi, M., University College London
Lettieri, P., University College London
Chapman, C., Advanced Plasma Power
Taylor, R.,
Originally presented on: 11/6/2013 16:15:00 - 16:35:00

## PERFORMANCE ANALYSIS OF RDF GASIFICATION IN A TWO STAGE FLUID BED - PLASMA PROCESS

M. MATERAZZI* **, P. LETTIERI*, R. TAYLOR **, C. CHAPMAN **

** Department of Chemical Engineering, University College London, Torrington Place, London WC1E 7JE, UK*

*** Advanced Plasma Power, South Marston Business Park, Swindon, SN3 4DE, UK*

Tar generation and ash disposal represent the strongest barrier for use of single stage fluid bed gasification for waste treatment, whereas sufficing for both is only possible with expensive cleaning systems and further processing. Waste fuels have, in general, physical and chemical characteristics much different from fossil fuels, so that the behaviour of these two feedstock types in high temperature fluid beds is very different one from the other. The high content of ashes and inorganic materials in RDF (Refuse Derived Fuel) dictates the thermal conditions in the fluid bed. A relatively low temperature has to be used to prevent agglomeration and sintering of bed material. For this reason, the gas that is produced by a standard fluid bed gasifier (FBG) has tars and other condensable organic species that are technically difficult and costly to remove. Furthermore, the bottom ash that is generated in the reactor may contain high levels of carbon, heavy metals and organic pollutants which lower the conversion efficiency of the process and limit any secondary usage.

The use of plasma systems has increasingly been applied with thermal waste treatment for its ability to completely decompose the input waste material into a tar-free synthetic gas and an inert, environmentally stable, vitreous material known as slag. The principal advantages that plasma offers to thermal conversion processes, besides the already mentioned tar/ash related issues absence, are a smaller installation size for a given waste throughput, and the use of electricity as energy source, characteristics which permit the technology to treat a wide range of low calorific value materials including liquids and solids. In applying the plasma technology to the gaseous products from a FBG, an advanced two-stage thermal process is able to achieve efficient conversion of the chars and the complex organics to the primary syngas constituents whilst limiting the electrical energy demand of the process (Figure 1).

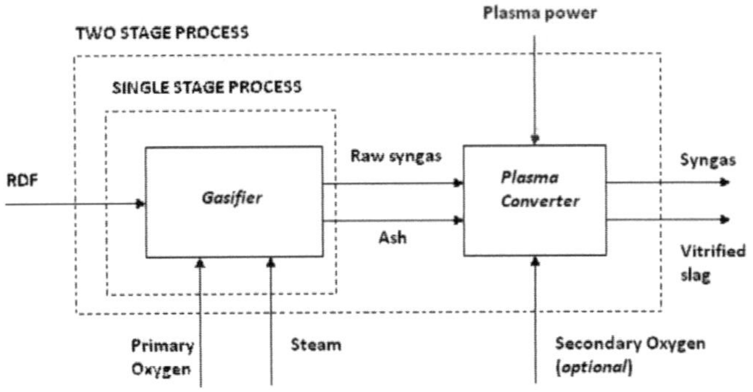


Figure 1: Schematic of Gasplasma two-stage thermal process.

This study focused on the thermodynamic assets of using a two-stage (FBG + plasma) thermal process over the conventional single-stage approach.Â  These include, for example, the fact that the primary thermal waste decomposition is performed in conditions of optimal stoichiometric ratio for the gasification reactants. Furthermore, staging the oxidant injection in two separate intakes significantly improves the efficiency of the system, reducing the plasma power consumption. The aim of the present work is also to determine the effects of the thermal parameters so as to be able to select optimum gasification conditions in a two stage process and to clarify the mechanism of thermal decomposition of carbonaceous materials in the FBG and in the plasma converter.

A flexible model capable of providing reliable quantitative predictions of product yield and composition after the two-stage process has been developed. The method has a systematic structure that embraces atom conservation principles and equilibrium calculation routines, considering all the conversion stages that lead from the initial waste feed to final products. The model was also validated with experimental data from a demonstration plant in Swindon (UK) (Table 1).

| | **Case 1** | **Case 2** | **Case 3** | **Case 4** | **Case 5** |
|---|---|---|---|---|---|
| **Description:** | | | | | |
| O2/fuel ratio (w/w) | 0.51 | 0.59 | 0.59 | 0.79 | 0.50 |
| Bed temperature (Â°C) | 770 | 720 | 795 | 720 | 800 |
| *Proximate analysis, % (w/w)* | | | | | |
| Fixed carbon | 6.4 | 12.2 | 11.6 | 8.5 | 22.8 |
| Volatile matter | 59.6 | 50.2 | 64.8 | 47.6 | 68.0 |
| Ash | 19.1 | 23.2 | 12.1 | 8.9 | 0.5 |
| Moisture | 14.9 | 14.4 | 11.5 | 35.0 | 8.7 |
| *Ultimate analysis,Â % (w/w)* | | | | | |
| C | 41.0 | 47.0 | 43.0 | 31.5 | 45.2 |

| | | | | | |
|---|---|---|---|---|---|
| H | 5.7 | 6.3 | 5.6 | 4.1 | 6.46 |
| O | 17.5 | 6.9 | 26.6 | 19.7 | 45.38 |
| N | 1.2 | 1.74 | 0.61 | 0.4 | 0.26 |
| S | 0.2 | 0.15 | 0.25 | 0.17 | 0.01 |
| Cl | 0.4 | 0.31 | 0.34 | 0.23 | 0.25 |
| GCV,Â  MJ/kg (dry basis) | 22.1 | 26.4 | 21.0 | 21.0 | 22.0 |

Table 1: Experimental parameters and characteristics of solid wastes (as received).

The study effectively demonstrated that the two-stage gasification system significantly improves the gas yield of the system and the carbon conversion efficiency, which are crucial in other single stage FBG systems, whilst maintaining high energy performances.

The most important outcome of this analysis is that the important parameters that define the performance of the fluid bed-plasma process are the oxidants (oxygen and/or steam) streams and the plasma power. These parameters are function of the type of waste and the detailed aim of the project. Furthermore, the two-stage performance can be quite accurately predicted by assumption of thermodynamic equilibrium.

Figure 2: Comparison of model results with experiment data from the (left) FBG gasifier; (right) FBG + Plasma converter

There is close correlation between the observed and predicted values of the syngas exiting the plasma converter, although the actual level of CO is still ~1% below the theoretical value while the $CO_2$ level is 1.5% higher. The noted differences are likely due to flow and RDF composition uncertainties rather than from the predictive model. Given the size and configuration of the demonstration plant and the inherent variability of real RDF uncertainties are to be expected. The layout of syngas heating value (LHV) and cold gas efficiency (CGE) have been calculated and analyzed. It was found out that high LHV and CGE values are maintained in different power and oxygen conditions. The reason is that addition of plasma power into the converter decreases the amount of secondary oxygen required for complete gasification and produces larger amounts of CO and $H_2$ in the product gas. The optimizing direction for the two-stage process can only be determined after considering the detailed situation on different projects.

**(534e) Process Modeling of Continuous Catalytic Gasification As a Waste to Energy Alternative**

Tillie, C. F., Cleveland State University
Riedthaler, M. J., Cleveland State University
Reeves, S. A., Cleveland State University
Adhikari, S., Cleveland State University
Gatica, J. E., Cleveland State University
Originally presented on: 11/6/2013 16:35:00 - 16:55:00

Presently, space exploration waste is either carried onboard for the duration of the mission or discarded into a trash module which is sent back into Earth's atmosphere for destruction. On the other hand, with an estimated cost of $10,000 per pound, the cost of delivering mass to locations beyond low earth orbit (LEO) locations is prohibitive.

Since approximately 15 pounds of waste are produced daily by a crew of four during a mission, the intrinsic value of this waste can be greatly enhanced by waste to energy technologies. Moreover, conceptualizing waste gasification as an in-situ alternative for processing of trash is also an efficient route to control waste, while maintaining a healthy habitable environment, during long-duration missions (Santiago-Maldonado et al., 2010).

The gasification of long-chain polymers is hypothesized to follow a mechanism of four combined reactions —two parallel oxidation reactions which are kinetically controlled, accompanied by two series reaction where oxidation products (carbon dioxide and carbon monoxide) are converted into methane and hydrogen. Integrating this process into spaceflight missions would greatly reduce the onboard mass requirements, as these products can be used as propellant fuel. Further, under certain conditions the process could be also be adapted to produce oxygen and water, two essential life support species for space missions.

This paper explores scale-up alternatives to implement low-temperature continuous gasification processes as potential waste management processes. A kinetic model formulated from batch gasification experiments is integrated into a process simulator along with a detailed thermodynamic description.

Emphasis is placed on the preparatory treatment of the waste slurries, the reactor module, and equilibrium stages for recycling reactants and recovering the catalyst. Detailed characterization of low and mid fidelity simulants is presented.

**(534f) Performance Evaluation of Accelerated Carbonation of Basic Oxygen Furnace Slag Via a Rotating Packed Bed: Modeling, Analysis and Maximization**

Pan, S. Y., National Taiwan University
Chiang, P. C., National Taiwan University
Chen, Y. H., National Taipei University of Technology
Chang, E. E., Taipei Medical University
Tan, C. S., National Tsing Hua University
Originally presented on: 11/6/2013 16:55:00 - 17:15:00

Performance of accelerated carbonation by basic oxygen furnace slag (BOFS) in a rotating packed bed (RPB) was evaluated in this investigation. The design of experiments (DOE) was introduced to evaluate the effect of different operating parameters including reaction time, particle size of BOFS, temperature, liquid-to-solid (L/S) ratio, and rotating speed on carbonation conversion of BOFS. A quadratic response surface was established statistically to yield maximum carbonation conversion based on experimental data. In addition, both the fresh and carbonated BOFS were characterized by the thermogravimetry and differential scanning calorimetry (TG/DSC), X-ray diffraction (XRD), and scanning electron microscopy equipped with X-ray energy dispersive spectroscopy (SEM-XEDS) to confirm the results of accelerated carbonation of BOFS in an RPB.

Furthermore, by combining the process chemistry, reaction kinetics, and mass transfer, a prediction model for carbonation of BOFS within an RPB, with an error less than 10%, was well-developed. The enhancement factors for mass transfer coefficient in case of carbonation of BOFS within an RPB were determined accordingly. In comparison with the developed model and the RSM results, a standard operation procedure (SOP) was proposed for maximizing the carbonation conversion of BOFS. It was thus concluded that the accelerated carbonation of BOFS using an RPB is a promising process for $CO_2$ capture due to its higher carbonation conversion achieving in a short reaction time.

**(534g) Production of Alumina From Fly Ash By a Mild and Cleaner Hydrometallurgical Process**

Zhong, L., Institute of Process Engineering, Chinese Academy of Sciences
Duan, D., Institute of Process Engineering, Chinese Academy of Sciences
Zhou, E.,
Han, H.,
Originally presented on: 11/6/2013 17:15:00 - 17:35:00

Fly ash is a large volume solid waste generated during the coal combustion by coal-fired power plants, wherein the valuable alumina content could be up to 30-50%. A mild hydrometallurgical process to produce alumina from fly ash only by alkaline solution at comparatively low pressure and temperature was investigated. This novel process is also a cleaner process, where the excess alkaline solution reactant is recycled to the leaching of fly ash, followed by the crystallizing of sodium aluminate hydrate and then its seed decomposing to prepare aluminum hydroxide, calcined to produce alumina. By controlling the phase transformation of the multiphase complex components in the fly ash, overall more than 90% of alumina was recovered from fly ash. The final residues could be utilized as feedstock in construction materials. Furthermore, the reaction mechanism of the process is investigated. This mild and cleaner process could also be appropriate for the produce of alumina from other high Al-containing solid wastes.

**(564a) Strategies in the Sustainable Design of Active Pharmaceutical Ingredients for Solvent Recovery Systems**

Savelski, M. J., Rowan University
Slater, C. S., Rowan University
Cavanagh, E. J., Rowan University
Originally presented on: 11/6/2013 15:15:00 - 15:40:00

The Chemical Engineering Department at Rowan University has partnered with several pharmaceutical companies and the US EPA to explore more sustainable pathways in the manufacturing of active ingredients and to educate future generations of engineers and decision makers in the importance of green engineering in pharmaceutical syntheses. The conducted research has evaluated greener designs for solvent recovery targeting not only the reduction of the overall solvent consumption, but the associated waste, and life cycle emissions from cradle to gate. These studies have explored greener separation processes and greener solvents using both LCA and economic metrics. Software has been developed to guide decision making and optimize processes based on these desired outcomes. When integrated correctly, solvent recovery is a viable option to reduce significantly the carbon footprint of drug synthesis as well as operating costs. Life cycle analysis of pharmaceutical manufacturing routes involving solvents such as acetonitrile, isopropanol, methanol, tetrahydrofuran, and methyltetrahydrofuran have been conducted. This paper will present the approaches taken to determine the optimal application of greener technologies to minimize emissions and make pharmaceutical manufacturing more sustainable.

**(564b) Towards a Complete Solution of Allocation in Life Cycle Inventories**

Hanes, R., The Ohio State University
Cruze, N.,
Goel, P. K., The Ohio State University
Bakshi, B. R., The Ohio State University
Originally presented on: 11/6/2013 15:40:00 - 16:05:00

The need for allocation arises in life cycle inventories when a process results in multiple products. In such a situation, resources and emissions embodied in the inputs need to be partitioned between the products. Popular approaches for accomplishing this task involve partitioning the embodied resources in proportion to characteristics of the outputs such as mass, energy, exergy or cost. Given the subjectiveness of this decision, many practitioners apply multiple allocation methods. Recent efforts have proposed the use of regression methods such as ordinary least squares and total least squares to solve the allocation problem. Allocation may be avoided by expanding the system boundary, but this approach can be tedious due to the need for more data.

We show that allocation can be understood as an ill-posed problem, analogous to the situation where the number of variables in simultaneous equations is larger than the number of independent equations. In such a situation, the number of solutions is infinite, and without additional information or assumptions, a single "correct" solution cannot be determined. Using a matrix framework for life cycle inventories, we devise a method of calculation whereby results from all possible allocation methods can be examined.

We use this insight to evaluate the special case where LCA is used to compare products that have the same use but are produced by different processes. Examples of such a case include comparing ethanol from various sources, gasoline, and other transportation fuels. We develop an approach to determine the sensitivity of the comparative LCA to allocation weights and show that sometimes the better product may be independent of the allocation approach. For cases when this does not happen, we propose an optimization approach to determine the "tipping curve" or the boundary across which the product preference changes.

The practical usefulness of this approach is demonstrated with two case studies. The first study compares the LCIs of 1,3-propanediol (PDO) produced from fossil fuels and from biomass. This comparison was found to be robust to allocation for nine resources and emissions out of the eleven considered: biomass PDO was preferred over fossil PDO for all choices of allocation weights. The second case study compared ethanol produced from corn stover hydrolysis, corn stover gasification, and corn grain fermentation. This comparison was not robust to allocation, although the two corn stover processes were preferred over the corn grain process for most choices of allocation weights. More information is needed to decide conclusively which process is preferred. Extension of the framework to incorporate this additional information is proposed.

**(564c) Hybrid Techno-Economic Modeling Tool for Greener Chemicals Supply Chains**

Yao, Y., Northwestern University
Chang, Y., Northwestern University
Thwaites, F., Northwestern University
Masanet, E., Northwestern University
Originally presented on: 11/6/2013 16:05:00 - 16:30:00

The demand of chemical products is rapidly increasing due to the growth of global economy. At the same time, the chemicals industry is a major source of global greenhouse gas emissions (GHG) because of its high energy consumption(1). As such, reducing the energy and emissions footprints of chemicals is a critical priority for environmentalists, manufacturers and policy makers. Promoted by this question, green supply chain management is receiving increasing attention because it can help chemical industry to reduce the overall environmental intensity and thereby offer "greener" products. Most supply chain initiatives and related studies apply process-based life-cycle assessment (LCA) techniques to model and quantify the environmental impacts and sustainability performance of chemicals(2). These LCA studies help people to understand the current environmental footprints of many chemicals, but they are not able to answer the question: how and where can we reduce the life-cycle energy and pollution footprints of these chemicals? This question requires the support of techno-economic technology potentials analysis, which is a vibrant facilities engineering and analysis field that has operated largely independent of LCA to date(3, 4).

A comprehensive decision support tool incorporating techno-economic analysis methods and data into the life cycle assessment (LCA) framework is presented in this paper. The techno-economic analysis includes detailed characterizations of discrete technologies for reducing energy use and environmental impacts, their remaining installation potential in various plants and economic sectors, and their relative cost and environmental performance when deployed in combination. With this analysis, the model can identify and incentivize cost effective improvements for green supply chains within the chemicals industry to produce maximal energy reduction at minimal environmental burden. It can provide chemical manufacturers with comprehensive understanding of the greatest opportunities and potential cost requirements for supply-chain energy savings and environmental improvements. Also, the adoption of new process and material can be easily incorporated into this model, which can help designers to prescreen the sustainability performance of their design on a higher level. This tool also provides policy makers with cost-effective and efficient policy options to promote the development of more sustainable chemical product supply chains.

References

1. ICCA. 2009. *Innovations for Greenhouse Gas Reductions* ICCA

2. Goldberg A, Holdaway E, Reinaud J, O'Keeffe S. 2012. *Promoting Energy Savings and GHG Mitigation Through Industrial Supply Chain Initiatives* Institute for Industrial Productivity

3. Masanet E, Matthews HS, Carlson D, Horvath A. 2012. *Retail Climate Change Mitigation: Life-cycle Emission and Energy Efficiency Labels and Standards : Final Report*: University of California, Berkeley

4. McKinsey & Company. 2009. *Unlocking Energy Efficiency in the U.S. Economy.*, McKinsey Global Energy and Materials New York, New York

**(564d) Design Decision-Making Under Uncertainty to Identify Sustainable Integrated Forest Biorefinery Strategies**

Sanaei, S., Ecole Polytechnique de Montreal
Stuart, P. R., Ã‰cole Polytechnique de MontrÃ©al
Originally presented on: 11/6/2013 16:30:00 - 16:55:00

Integrated biorefinery into existing pulp and paper facilities is a potential game-changing solution for the transformation of North American forestry industry, and the manufacture of an expanded product portfolio. However not all biorefinery strategies are sustainable and not have the same level of risk. Thus it is crucial that unsustainable strategies be systematically screened-out from the list of possibilities at the early design stage. The methodology presented in this study assesses candidate biorefinery alternatives from economic, competitiveness and environmental perspectives using different process systems analysis tools (e.g, techno-economics, Life Cycle Analysis, etc.) for both deterministic and uncertain conditions. The results of these assessments are used to evaluate a set of sustainability criteria, despite of the conventional analyses which generally use only short-term profitability metrics for decision-making. These conflicting criteria were aggregated into a unique sustainability score for each candidate biorefinery strategy by conducting a multi-criteria decision-making (MCDM) panel by which candidate strategies are ranked using a sustainability perspective.

However the candidate biorefinery strategies imply different levels of technology and market risk due to different sources of uncertainties which should be addressed in decision making. Thus to make a realistic strategic decision, the sustainability of biorefinery strategies should be assessed under uncertainty. In this study, different sources of uncertainty including market, technology and panel uncertainties are considered in biorefinery decision-making. In order to address this objective, a stochastic risk analysis method (Monte-Carlo Analysis) was used to develop probability distribution functions for sustainability criteria, which were then employed in a second MCDM panel. The proposed methodology in this study not only takes into account a complete set of sources of uncertainty, but also reflects the risk attitude of decision-makers in utility functions applying risk aversion theory. Comparing the results of deterministic and uncertain conditions shows that by evaluating uncertainty and quantifying risk attitude of panelists, the basis for the panel decision changes. This confirms the necessity of systematically addressing risk in biorefinery strategic decision making.

**(564e) MILP-Based Clustering Method for Multi-Objective Optimization: Application to Environmental Problems**

Oliva, D. G., Universitat Rovira i Virgili
Guillen-Gosalbez, G., Universitat Politicnica de Catalunya
Jiminez-Esteller, L., Universitat Rovira i Virgili
Originally presented on: 11/6/2013 16:55:00 - 17:20:00

Multi-objective optimization (MOO) has recently emerged as a useful technique in environmental engineering. One major limitation of this approach is that its computational burden grows rapidly with the number of environmental objectives, which causes difficulties regarding the computation and visualization of the Pareto solutions. In this work, we present several theoretical and algorithmic developments for grouping environmental objectives into clusters on the basis of which the multi-objective optimization can be performed, thereby facilitating the computation and analysis of the Pareto solutions. Particularly, we study how to group objectives of a MOO problem into meaningful clusters with the property that the minimization of any objective within a cluster will result in the minimization of the rest of objectives within the same cluster.

Our method is based on a novel mixed-integer linear program (MILP) that identifies in a systematic manner groups of objectives that behave similarly. The capabilities of the approach presented are tested through its application to several case studies, in which we compare the results of the MILP with those produced by a statistical method. The statistical method consists of a standard clustering algorithm.

The first case study deals with the design of petrochemical supply chains considering several environmental indicators simultaneously (Pozo et al., 2012). The second example deals with the design of bio-ethanol supply chains in Argentina considering 6 objectives (Kostin et al., 2012). The third example addresses the multi-objective optimization of hydrogen supply chains for vehicle use in Spain considering several Life Cycle Assessment impacts (Sabio et al., 2012). In the last example, we study an artificial dataset with 10 objectives and 41 points generated randomly. The goal is to test the performance of the statistical method when the objectives are poorly correlated.

Numerical results show that when the objectives are highly correlated, both methods (i.e., the MILP and the statistical method) lead to similar results, being the statistical approach the one with lower CPU times. In contrast, the statistical method leads to clusters with larger errors that differ from those produced by the rigorous MILP approach when the correlation between objectives is low. Our rigorous MILP-based approach is the only one that guarantees the minimum approximation error.

## (564f) Emergy Method Analysis On Sludge Residues For Use From Waste-WATER Treatment Plant

Cano, N. A., Universidad Nacional de colombia
Gallego, D.,
VelÃ¡squez, H. I.,
Originally presented on: 11/6/2013 17:20:00 - 17:45:00

## EMERGY METHOD ANALYSIS ON SLUDGE RESIDUES FOR USE FROM WASTE-WATER TREATMENT PLANT

### ABSTRACT:

It was confirmed that the emergy analysis method allows comparison between the two alternatives for biosolids disposal while integrating disposition the energy, economic and environmental aspects. Therefore, by using the method of emergy analysis as a tool for sustainability assessment for environmental decision making, and understanding the sustainability of a disposal process that would reach a greater economic benefit with minimal environmental load, the actual benefit of biosolids "using biosolids as Soil Fertilizer" was evaluated and compared against a suitable alternative referenced in the literature as "energy recovery from Biosolids using fluidized bed gasification" under two scenarios: biosolids considered as a renewable resource and as an imported resource.

Being the best alternative according to the sustainability indices "Energy Recovery from Biosolids Using Fluidized Bed Gasification", ie, this alternative is energetically profitable, economically viable and environmentally acceptable; It does not mean that the actual use of the biosolids is economically or environmentally inappropriate. On the contrary, it does not produce environmental stress because its economic performance is not being taken advantage of at its full potential.

The analysis of emergetic levels not only allow us to evaluate the sustainability of a project, It also offers the possibility to improve the practice, so for example for the current use of biosolids the sustainability index allows us to say that a pretreatment of sewage sludge before its disposal improve to the alternative and become the most acceptable [1]

**KEYWORDS:** Emergy, sustainability, indicators, waste water, disposal, sludge

In Colombia, it is not uncommon to rely on plants for waste-water treatment purposes followed by the anaerobic stabilizing of biosolids as an alternative source of treatment. This practice is due to the many topographical limitations and conditions that do not allow mainstream treatment such as the use of oxidation ponds.

However, during waste-water treatment processing, large significative quantities of highly putrescible bio-solids with a higher energy concentration as well as organic charge/volume are produced; such is the case-example of the waste water treatment plant in San Fernando (an EEPPM-owned plant) where more than 70 tons of bio-solid material are generated, to be used to treat the 1.100m l/s of water needed to be released on a permanent basis [2].

Although bio-solid materials processed in this particular example is known for carrying similar traits to those found in urban areas, as well as being defined as neither toxic nor hazardous, they do in fact carry contaminants that required treatment and profitable use of such. Based on these findings, many alternative traditional methods for treatment and use of bio-solids have been implemented, those being for energy production, materials for construction purposes, restoration of natural habitat as well as agricultural uses [3]. However they have proven to be ineffective in their result.

In retrospect; it can be said that management of waste-water treatment is ultimately lacking and inconclusive, lest treatment of its additional waste product is carried-on in a sustainable fashion. defining sustainability of course, as the process of obtaining the highest financial return with a minimum effect in the environment; for such reason, the use of the emergy analysis method is a reliable gauging tool to measure level of sustainability to take full advantage of such, as proposed by [4] [5] [6] [7], as the most prominent and noteworthy method used in other countries [8] [9] [10] [11] [12] for sustainability based practices, in a manageable and equitative way in the economic, energy and environmental areas.

The emergy analysis approach is known as an analytical method, that through its integration of economy, ecology and thermodynamics concepts allows the assessment of economic, social and environmental impacts in a theoretical form, on development-driven projects, thus becoming a stronghold system in the fields ecological economy and environmental engineering.

As an emergy analysis method-based research, its ultimate intent is to weigh-in the efficient level of sustainability of any processing system used to make decisions concerning the environment. this method will be used in a way whereby a correct approach to evaluate all sludge residues found in any waste-water treatment plant can be identified in order to choose an assessment alternative that is ′applicable′. that is to say, an alternative that is ′′energy-suitable′′, ′′financially viable′′, and ′′environmentally friendly′′, in accordance to sustainability indicators; this research means to assess the use of the emergy analysis method as a sustainability appraisement tool of bio-solids production at a waste-water treatment plant in Colombia, while being able to satisfy all economic, environmental and energy levels of sustainability. Having as contrast, an already existing method of assessment found in the table of reference, allowing evidential proof of its utility on projects of a similar nature, as well as measuring if current alternative methods of waste-water treatment used in Colombia fit the standards of sustainability, by using the following parameters.

In order to meet such general objective, the application and results of the emergy analysis method used in a waste-water treatment plant in Colombia is analyzed, followed by an analysis of the system of sustainability. Then, by comparing the assessment of the two chosen alternatives of bio-solids used, through the estimate of emergy of both alternatives, the better method for bio-solid analysis is defined and accepted as the best one.

Thus, based on the sustainability indicators, the recovery of bio-solids energy through fluidized bed gasification becomes the best alternative for sustainability. That is to say, this alternative, meets all three standards of sustainability when compare to the alternative of using bio-solids as ground fertilizer.

## REFERENCES

[1] Cano Londoño N. A. (2012). Análisis mediante el matodo emergetico de la disposición de los

lodos producidos en una planta de tratamiento de aguas residuales. (Aplicación a una PTAR en el Área Metropolitana del Valle de Aburrá). [Tesis MSc.]. Universidad Nacional de Colombia Sede Medellín.

[2] Velez, J. (2007). Biosolids: A problem or a solution? *Producción mas Limpia*, *2*, 57-71.

[3] Fytili, D., & Zabaniotou, A. (2008). Utilization of sewage sludge in EU: application of old and new methods—A review. *Renewable and Sustainable Energy Reviews*, *12*(1), 116-140.

[4] Agostinho, F., Diniz, G., Siche, R., & Ortega, E. (2008). The use of emergy assessment and the Geographical Information System in the diagnosis of small family farms in Brazil. *Ecological Modelling*, *210*(1-2), 37-57.

[5] Cherubini, F., Bargigli, S., & Ulgiati, S. (2009). Life Cycle Assessment (LCA) of waste management strategies: Landfilling, sorting plant and incineration. *Energy*, *34*(12), 2116-2123.

[6] Cohen, M. J., Brown, M. T., & Shepherd, K. D. (2006). Estimating the environmental costs of soil erosion at multiple scales in Kenya using emergy synthesis. *Agriculture, Ecosystems & Environment*, *114*(2-4), 249-269.

[7] Odum, H., & Odum, B. (2003). Concepts and methods of ecological engineering. *Ecological Engineering*, *20*(5), 339-361.

[8] Geber, U., & Bjo, J. (2001). The relationship between ecosystem services and purchased input in Swedish wastewater treatment systems — A case study. *Ecological Engineering, 18, 39-59.*

[9] Liu, Y., Kong, S., Li, Y., & Zeng, H. (2009). Novel technology for sewage sludge utilization: preparation of amino acids chelated trace elements (AACTE) fertilizer. *Journal of hazardous materials*, *171*(1-3), 1159-67.

[10] Siracusa, G., & La Rosa, A. D. (2006). Design of a constructed wetland for wastewater treatment in a Sicilian town and environmental evaluation using the emergy analysis. *Ecological Modelling*, *197*(3-4), 490-497.

[11] Zhou, J. B., Jiang, M. M., Chen, B., & Chen, G. Q. (2009). Emergy evaluations for constructed wetland and conventional wastewater treatments. *Communications in Nonlinear Science and Numerical Simulation*, *14*(4), 1781-1789.

[12] Zhou, S. Y., Zhang, B., & Cai, Z. F. (2010). Emergy analysis of a farm biogas project in China: A biophysical perspective of agricultural ecological engineering. *Communications in Nonlinear Science and Numerical Simulation*, *15*(5), 1408-1418.

**(575a) Direct Land Use Change Greenhouse Gas Emissions Associated With Forest-Based Biofuels and Bioenergy Production in Michigan**

Fan, J., Michigan Technological University
Froese, R. E., Michigan Technological University
Shonnard, D. R., Michigan Technological University
Handler, R. M., Michigan Technological University
Originally presented on: 11/6/2013 15:15:00 - 15:37:00

Michigan has one of the largest amounts of timber net growth in excess of removals of any state (135.6 million cubic feet per year). Some of this excess timber growth could be a promising feedstock to address the state's forest products industry while also creating a thriving renewable fuels industry. However, harvesting of forest resources for bioenergy production could reduce terrestrial C storage in the near term. Therefore, forest-based bioenergy LCA should incorporate forest carbon stock assessment over a long time frame to estimate the total greenhouse gas (GHG) emissions. In this study, the Carbon Budget Model of the Canadian Forest Sector (CBM-CFS3) is applied to simulate the C stock change in Michigan over a greater than 200 year time frame for aspen-dominated forests, where aspen is intensively harvested for biofuel and bioenergy production. Under intensive harvesting, additional aspen biomass could produce an extra of 18.46 billion gallons of ethanol to blend with gasoline for the transport sector over the next 250 years. Alternatively, the same biomass could be used to produce 32.25 billion gallons of bio-oil by fast pyrolysis process, which can be combusted to generate electricity or upgraded to renewable gasoline and diesel. Model results show that the biofuels (ethanol or pyrolysis oil) and bioenergy (electricity) result in higher GHG emissions than using fossil fuels and business-as-usual aspen harvesting due to the forest carbon loss in the short term. However, the intensive management and harvest of aspen helps accumulates more C in the ecosystem over the long term, which results in negative dLUC emissions (credit) eventually. The time required for this initial carbon debt to be repaid is approximately 60 years for both biofuels; thereafter, this aspen harvesting and management approach accrues GHG savings compared to use of fossil fuels. This study also shows that forest carbon dynamics and dLUC emissions are highly sensitive to the assumed post-harvest growth curve, which can change the time required to overcome the initial C debt.

**(575b) Light-Limited Continuous Culture of Micro Algae in a Taylor Vortex Reactor**

Vigil, R. D., Iowa State University
Kong, B., Iowa State University
Originally presented on: 11/6/2013 15:37:00 - 15:59:00

Recently, it has been demonstrated that Taylor vortices – hydrodynamic structures that arise in the annular region between two concentric cylinders when the inner cylinder rotates – can substantially improve the growth rate of algal biomass in a batch photobioreactor by inducing the flashing light effect. In order to assess the potential for using Taylor vortex flow to continuously culture algae, experiments were carried out in a continuous flow Taylor vortex algal photobioreactor using various microalgae. Two important operating parameters were varied: the dilution rate and the inner cylinder rotation speed. For a fixed inner cylinder rotation speed, biomass productivity decreased at low dilution rates but was constant at higher dilution rates. Biomass productivity was found to increase slightly with increasing inner cylinder rotation speed for a fixed dilution rate, and this weak dependence on rotation speed may indicate that the flashing light effect was nearly maximized even at the lowest cylinder rotation speeds employed. Overall, it is demonstrated that a continuous flow Taylor vortex algal photobioreactor can be used to produce and sustain high biomass production and carbon dioxide capture rates when operated in continuous flow mode.

**(575c) Algae Derived Bio-Oil Separation By Supercritical Fluid Fractionation**

Ginosar, D. M., Idaho National Laboratory
Agblevor, F., Utah State University
Petkovic, L. M., Idaho National Laboratory
Newby, D. T., Idaho National Laboratory
Moses, J. M., CF Technologies, Inc.
Originally presented on: 11/6/2013 15:59:00 - 16:21:00

Bio-oils produced by thermal processes, such as fast pyrolysis, catalytic pyrolysis or hydrothermal liquefaction, are a promising source of sustainable, low greenhouse gas (GHG) alternative fuels. These thermal treatment processes are also well suited to decentralized energy production due to low capital and operating costs. However, pyrolysis bio-oils from algae have complex chemical compositions, including large concentrations of water, carboxylic acids, carbohydrates, and protein-derived substances. The oils are viscous, reactive, and thermally unstable. Due to the occurrence of decomposition products together with water and an abundance of compounds of many classes, range of polarities, and the large fraction of oxygenated compounds the mixture is difficult to fractionate by distillation, chromatography, or membrane based processes.

Several methods have been explored to separate bio-oils into stable fractions. By far the most effective separation and stabilization method has been solvent extraction. Although effective, solvent extraction poses two main obstacles to commercialization: (1) the significant amounts of energy required to remove the solvent from the product and (2) the propensity of the solvent to be contaminated with minerals from the char or ash. Separation of thermally produced bio-oils using supercritical fluids offers the advantages of liquid solvent extraction while drastically reducing energy demands and the predisposition to carry over solids into the extracted phase.

*Scenedesmus dimorphus* was pyrolized at 350, 400, and 450°C with and without a catalyst. Near critical and supercritical carbon dioxide and propane were evaluated for the extraction and fractionation of algae pyrolysis oils into stable fractions. Extractions were carried out at 25, 40, 65 and 110°C at a reduced pressure of 2.0. The effect of pyrolysis conditions and extraction conditions on product composition and stability will be discussed.

**(575d) Enhanced Separation of Algal Biomass By Vibratory Shear Microfiltration for the Sustainable Production of Biofuels**

Slater, C. S., Rowan University
Savelski, M. J., Rowan University
Kostetskyy, P., Rowan University
Originally presented on: 11/6/2013 16:21:00 - 16:43:00

The dewatering step in the production of algal biofuels is energy intensive resulting in a significant carbon footprint. Membrane separations are one of the processes being investigated to increase efficiency of the operation, but have limitations due to concentration polarization and fouling. This is a concern that limits the scale-up potential of this process for commercial biorefineries. To overcome issues with flux degradation, we have investigated shear-enhanced cross-flow membrane filtration processes. High membrane shear rates were achieved by high-frequency torsional oscillations of the membrane unit. Operated at the resonant frequency, these systems can provide enhanced separation performance at minimal power inputs. Several microfiltration membranes have been evaluated for the separation of dissolved *Chlorella Vulgaris*, a potential feedstock for biodiesel production. This process provides a greener alternative to conventional systems, as high values of flux and separation efficiency can be maintained and the biomass can be effectively concentrated. Studies to evaluate the effect of algae feed concentration, membrane pore size, trans-membrane pressure, and vibrational amplitude on permeate flux were performed. The integration of vibratory field membrane separation into an overall design for the dewatering of algal biomass can result in a more sustainable, greener system.

**(575e) Multi-Objective Optimization of Biobutanol Production: Continuous Fermentation Coupled With Vacuum Separation Unit**

Sharif Rohani, A., University of Ottawa
Thibault, J., University of Ottawa
Mehrani, P., University of Ottawa
Originally presented on: 11/6/2013 16:43:00 - 17:05:00

Butanol produced via acetone-butanol-ethanol (ABE) fermentation of biomass, is a liquid biofuel that offers many advantages over other biofuels such as bioethanol and is targeted to partly replace gasoline. In ABE fermentation process, a strain of bacteria such as *Clostridium acetobutylicum*consumes sugar to produce ABE solvents. Due to the inhibition of butanol, ABE fermentation is known for its low solvent concentrations in the broth. Thus, to produce butanol more economically, it is required to operate the fermentation process under optimal conditions. Multi-objective optimization has been used to determine Pareto-optimal solutions where optimal decisions are made in the presence of two or more conflicting objectives. Based on an available kinetic model and a set of related mass balances for ABE fermentation, a continuous fermentation system coupled with a vacuum separation unit was simulated and a large number of Pareto-optimal solutions were obtained based on 4 decision variables (feed sugar concentration, dilution rate, evaporation rate and biomass concentration), and 3 objective functions (specific productivity, butanol concentration and sugar conversion) in which were maximized.

The optimal solutions were then ranked using the Net Flow Method and were compared to the continuous fermentation without the vacuum separation unit. It was found that for the best solution, the selective removal of solvents via the vacuum separation unit enhanced the specific productivity of the ABE fermenter by 51% for the same substrate conversion. The final average butanol concentration in the vacuum fermentation was 14.8 g/L which represents an increase of 70% compared to that in the standard continuous fermentation. Moreover, for both fermentation cases, it was shown that the sugar loss is extremely high which has an adverse effect on profitability of the process. In addition to sugar loss, butanol yields over sugar consumption for the best points in standard continuous system and continuous system coupled with vacuum were 10.1% and 13.9%, respectively.

**(575f) Growth, Harvesting, and Modeling of Lipomyces Starkeyi Lipids On Waste Sweet Potatoes**

Mason, M.,
Dufreche, S., University of Louisiana at Lafayette
Zappi, M. E., University of Louisiana at Lafayette
Subramanyam, R.,
Bajpai, R., University of Louisiana at Lafayette
Originally presented on: 11/6/2013 17:05:00 - 17:27:00

The production of fuels from renewable resources has traditionally faced many barriers, with the most common being limited feed stocks for utilization. The growing need for cost effective renewable resources is a driving force in the development of potential feed stocks. The sweet potato has shown in previous research to have a very high starch content. In addition, up to 20% of the annual sweet potato harvest is discarded due to unsuitable use as a food item. The yeast *Lipomyces Starkeyi* had previously shown the ability to produce lipids from pure starch, but not raw extract from whole sweet potatoes. The soluble starch obtained from this waste was used for the microbial conversion of starch to lipids utilizing a cheap feedstock. The soluble starch was evaluated for methods that would optimize production of the starch for conversion to lipids. Techniques were developed for preparing the sweet potatoes and maximizing extractable soluble starch. The starch obtained was then used as a feedstock for *Lipomyces starkeyi* to evaluate lipid production. Results to be presented include analysis of yeast growth and lipid accumulation, extraction efficiencies, and an economic analysis of full-scale lipid production from sweet potatoes.

**(575g) A New Approach to Study Selected Aspects of Biomass Oxidation in Supercritical Water**

Mohammad, J.,
Hegde, U. G.,
Hicks, M.,
Kozinski, J. A., University of Saskatchewan
Originally presented on: 11/6/2013 17:27:00 - 17:49:00

A variety of reactor systems are used in supercritical water oxidation (SCWO) including stirred reactors, transpiring wall reactors, flow through reactors and centrifugal reactors. Chemical reactions using SCWO technology take place in the thermodynamically favorable supercritical regime where high conversion rates and are not noticeably limited by heat transfer and mass transfer rates.

One of the advantages of using SCWO for biomass conversion is that supercritical water acts as an environmentally green solvent. Because biomass has an initial water content of 10-25 wt.%, SCWO technology is advantageous because the supercritical water serves as a solvent to breakdown the organic material prior to oxidation without the need for a drying step. However, conventional thermo-chemical techniques, such as pyrolysis and gasification utilized for the production of green fuels and chemicals, require the biomass to at least be partially dried.

During this study a portable SCWO reactor system, the SCWO Test Facility developed by NASA, has been utilized to study the conversion of beetle-infested wood into valuable fuels and chemicals. The SCWO Test Facility is originally designed for studying the use of SCWO for processing waste streams comprising plastics, papers and bio-wastes that would be generated during extended space missions. However, this unique system can also be used for processing aqueous waste streams containing small organic molecules such as methanol and acetic acid that would otherwise be difficult to be removed by conventional methods. A detailed study involving both biomass model compounds and the raw biomass has been undertaken to evaluate the system's overall conversion efficiency.

One of the salient features of the NASA SCWO Test Facility is that it can be operated as a constant volume batch reactor or a constant pressure continuous reactor. A piston accumulator is installed downstream to regulate the reactor gas pressure. Real-time in-situ measurements of the reactant and product concentrations are performed using the high temperature optical fiber and laser-assisted probe connected to a Raman spectrometer.

During this study the SCWO reactor temperatures and pressures have been varied in the range of 400-500 $^{0}$C and 3400-4400 psi by maintaining a constant feed/water ratio. The primary objective of this green technology is to recover synthetic gas ($H_2$–CO) as a gaseous fuel while maximizing the production of value-added chemicals such as levoglucosan, furfural, hydroxyacetaldehyde, acetic acid, and methanol. Both gas and liquid product yields are investigated, analyzed (GC-MS, NMR and FTIR) and presented.

**(600a) Mixed-Conductor Supported Ferrites for Chemical Looping Conversion of Methane**

Chen, Y.,
Li, F., North Carolina State University
Originally presented on: 11/7/2013 8:30:00 - 8:51:00

The chemical looping strategy represents a promising option for carbon dioxide capture from fossil fuels. The chemical looping processes utilize solid oxygen carriers to indirectly convert carbonaceous fuels to concentrated $CO_2$ and carbon free products via redox reactions. Due to its low cost and abundance, iron oxide is a desirable option as an oxygen carrier. The challenge for iron oxide based oxygen carriers resides in their low activity for methane and solid fuel conversions. Our recent studies have indicated that mixed ionic and electronic conducting (MIEC) oxides, when used as the "support", can significantly enhance the redox activity of ferrite based oxygen carriers. In the current study, ferrite based oxygen carriers supported with various MIEC perovskites are synthesized. The redox activities of these oxygen carriers are evaluated using a thermogravimetric analyzer (TGA) with $H_2$ and $CH_4$ as the reducing gases. The phase structures of these samples, before and after redox reactions, are determined using X-ray diffraction (XRD) and the morphologies of the samples are characterized by SEM. The results confirm the high redox activity of the MIEC supported ferrites. A general correlation between the mixed conductivity of the support and the redox activity of the oxygen carriers are observed. Long term redox reactions and microscopic/spectroscopic characterizations are also carried out to determine relationships between support composition/structure and the long term stability of the ferrite-based oxygen carriers.

**(600b) A Novel Mixed Metallic Ni Based Oxygen Carrier for Chemical Looping Combustion**

de Lasa, H., University of Western Ontario
Quddus, M. R., The University of Western Ontario
Hossain, M. M., King Fahd University Petroleum and Minerals
Originally presented on: 11/7/2013 8:51:00 - 9:13:00

This study reports the development of a high surface Ni based oxygen carrier (OC) and the investigation of its reactivity with gaseous fuel for Chemical-Looping Combustion (CLC). High surface area facilitates high dispersion of metal particle and increases the oxygen carrying capacity. On the other hand, high surface area also enhances the formation of irreducible species like $NiAl_2O_4$, which minimize the reactivity and oxygen transport capacity. Therefore, to achieve the high surface OC while preserving its reactivity, stability and fluidizability, a Ni based OC was developed in this study over Co promoted La modified $\gamma$-$Al_2O_3$ using incipient wetness impregnation technique. The $\gamma$-$Al_2O_3$ was first modified with addition of 5wt% La. The resulting La/$\gamma$-$Al_2O_3$ was then impregnated with 1wt% Co. Finally, a total of 20wt% Ni was loaded to the OC precursor via successive impregnations of 5wt% Ni. The resultant paste after each successive metal loading was reduced under three controlled gas flow rates containing 10% $H_2$ with the balance being He. The surface area was found to vary considerably based on the gas flow rates used during the preparation stages. 50-120 $m^2/g$ surface area was found for OCs prepared under low reducing gas flow. These synthesized OCs display excellent reducibility and stability in successive TPR/TPOs and $H_2$ chemisorptions. TPR profiles for the two OCs consistently show 95% reducibility at/below 750 °C. These results are encouraging given that they demonstrate the oxygen in the OC is very close to the expected amount. These results also show that the OC reducible phase is primarily NiO with small fractions of $NiAl_2O_4$. The controlled gas flow rates help the formation of easily reducible NiO species minimizing Ni-support interactions and formation of $NiAl_2O_4$. It was also found that the reduction behavior of these OCs resemble the chemical properties of $\delta$- and $\theta$-$Al_2O_3$ for low gas flow rates. In contrast, the structural integrity and characteristic chemical properties of Ni over $\gamma$-$Al_2O_3$ was observed in the case of high gas flow rates. Considerable amount of Ni was observed to be present as $NiAl_2O_4$ in this case. However, the absence of $\theta$- and/or $\delta$-phase of $Al_2O_3$ in the XRD experiments suggest that the phase transformation might occur at the interfaces between the Ni crystallites and alumina support surfaces, whereas the bulk structure of the support remained in the $\gamma$-phase. The SEM/EDX and $H_2$ pulse chemisorption results confirmed that all the three prepared OCs were highly stable in consecutive redox cycles. A well dispersed state of Ni crystals and the absence of agglomerations were observed for both the fresh and used OCs.

The reactivity of these prepared OCs was studied under the expected conditions (turbulent fluidized bed and high temperature) of an industrial scale CLC unit in CREC Riser Simulator. The CREC Riser Simulator is a bench scale mini-fluidized bed reactor capable of simulating circulating fluidized bed reactions conditions. The reactivity study was performed with stoichiometric amount of $CH_4$. The $CH_4$ conversion was found to be 81% ($A_{BET}$ = 50 $m^2/g$) and 86% ($A_{BET}$ = 120 $m^2/g$) for the OCs prepared under low gas flow rates. Maximum 88% $CH_4$ conversion was observed for the OC with high surface area ($A_{BET}$ = 140 $m^2/g$). The increased conversion of $CH_4$ was resulted from the increased oxygen transport capacity with the surface area of the carrier particles. However, the selectivity of $CO_2$ was found to decrease from 97% to 82 % with the increasing surface area.  This was mainly due to the increased catalytic effect resulting from highly dispersed Ni crystals over high surface area.

This systematic investigation of the reactivity of the Ni particles over wide range of crystal sizes reveals the fact that the low surface and low dispersed Ni based OC is favorable for CLC operation, whereas highly dispersed Ni based OC is suitable for CLR process. The oxygen transport capacity can be significantly increased with increasing surface area.

**(600c) Low-Cost Oxygen Carrier Materials for Chemical-Looping Combustion of Biomass**

Schwebel, G.,
Vilela Perez, G., Chalmers University of Technology
Knutsson, P., Chalmers University of Technology
Leion, H., Chalmers University of Technology
Originally presented on: 11/7/2013 9:13:00 - 9:34:00

Chemical-looping combustion (CLC) is one of the most promising methods to implement carbon capture technologies to mitigate $CO_2$emissions from thermochemical fuel conversion.

The solid oxygen carrier is a core component of every CLC system and the choice of the oxygen carrier depends on the fuel and operation conditions. When it comes to conversion of solid fuels, low-cost oxygen carriers seem to be more suitable. This is because the lifetime of an oxygen carrier in a CLC system may be limited by side reactions with fuel ash, or by carryover losses in the ash separation.

Hence the use of low cost oxygen carrier materials is preferred, lowering the economic and technologic effort connected to higher demands of make up.

Due to its carbon neutrality, the use of biomass in carbon capture strategies leads to net negative emissions. However, the ash of biomass is known to be more reactive and has generally lower softening and melting temperatures than most coals. This makes biomass combustion units more prone to defluidization, slagging and fouling and corrosion. If biomass should be used as fuel the temperature is usually kept at lower levels to overcome these problems.

In the present contribution different low cost oxygen carrier materials are investigated with respect to their reactivity at different temperatures. The data is obtained from experiments carried out in a quartz glas reactor under fluidized bed conditions. The bed temperature was set to 700, 800, 850 and 950°C, respectively. The perforance was tested with volatile components and in selected cases the influence of the char gasification has been investigated separately. Before the actual experiments, the oxygen carrier materials were activated at 850 and 950°C using a mixture of $H_2$ and CO as fuel.

**(600d) Fe and Mn-Based Oxygen Carriers for Chemical Looping Combustion**

Bhavsar, S., US DOE-National Energy Technology Laboratory, Pittsburgh
More, A.,
Veser, G., University of Pittsburgh
Originally presented on: 11/7/2013 9:34:00 - 9:56:00

Chemical looping combustion (CLC) is an emerging technology for clean energy-production from fossil and renewable fuels based on cyclic oxidation and reduction of an oxygen carrier. These carriers – typically metal oxides – have to show sufficient thermal and mechanical stability to withstand the demanding conditions of CLC while at the same time yielding high carrier and fuel conversion. Among the wide range of materials investigated, extensive work has recently been focused on iron and manganese oxides as oxygen carriers due to their abundance, low-cost, and environmentally benign nature. However, both metals show comparatively low reactivity and only moderate selectivity for total oxidation products.

In the present study, we investigated the use of Mn-Fe mixed-metal carriers for CLC. We had previously shown that doping Fe-based carriers with Ni can yield significant enhancements in reactivity. Ceria was chosen as support based on our previous findings that ceria can result in strong enhancements of carrier utilization and stability of the bimetallic metal phase.

Fe/Mn mixed oxides were synthesized with various Fe/Mn ratio. All carriers were synthesized with 40wt% metal by simple incipient wet impregnation technique on ceria supports. The carriers were characterized (as synthesized and post-reactive tests) using TEM, XRD, BET, and EDX. In a first step, the carriers were tested in a thermogravimetric analyzer (TGA) with $H_2$ and $CH_4$ as fuels. All FeMn carriers showed good reactivity and thermal stability with no signs of metal-support spinel formation. Increasing the Fe/Mn ratio resulted in improved oxygen carrying capacity beyond a simple linear addition of the capacity of the pure metal components. Next, selectivity and fuel conversion was further evaluated in fixed-bed reactor experiments with $CH_4$ as fuel. For both pure metals, a transition in selectivity was observed from total oxidation at high carrier oxidation states to partial oxidation over the reduced carriers. Interestingly, the bimetallic carriers showed a combination of the high syngas selectivity of Mn-based carriers with the widened operating window of Fe-based carriers, rendering this bimetallic carrier material of particular interest for partial oxidation of methane via chemical looping.

**(600e) Experimental and Model Comparison of Mixed Metal Ferrites for H2 Production Using Chemical Looping**

Aston, V. J., University of Colorado at Boulder
Evanko, B. W., University of Colorado at Boulder
Muhich, C. L., University of Colorado at Boulder
Scheffe, J., University of Colorado at Boulder
Weimer, A. W., University of Colorado at Boulder
Originally presented on: 11/7/2013 9:56:00 - 10:17:00

The chemical looping hydrogen (CLH) production process is an alternative method to water-gas shift to generate pure $H_2$, as well as a separate stream of pure and easily-sequestered $CO_2$, from any synthesis gas without complicated and expensive separations equipment. Synthesis gas is used to reduce a metal oxide material at moderate to high temperatures (500°C to 1000°C) as it leaves a gasifier, producing a stream of $CO_2$ and $H_2O$. The resulting reduced metal is then oxidized with $H_2O$ to produce $H_2$ and regenerate the metal oxide. In this study, the mixed metal ferrites $CoFe_2O_4$ and $NiFe_2O_4$ were investigated experimentally and shown to meet the criteria for CLH metal oxides. A comparison was made with $Fe_2O_3$, which is the current state of the art CLH metal oxide. At low solid conversions the mixed metal ferrites allowed the recovery of nearly three times more $H_2$ during $H_2O$ oxidation than was recovered using $Fe_2O_3$ under the same conditions. This $H_2$ recovery required additional steam generation due to the low $H_2O$ to $H_2$ conversion during the oxidation of near-stoichiometric $CoFe_2O_4$ and $NiFe_2O_4$. A system model developed based on the mixed metal ferrite equilibrium with the $H_2/H_2O$ and $CO/CO_2$ systems was used to investigate the effects of syngas composition, temperature and a supplementary $O_2$ oxidation on the steam production required for complete oxide regeneration and the overall efficiency of the system. The $CoFe_2O_4$ and $NiFe_2O_4$ materials offer significant benefit over conventional $Fe_2O_3$ at lower reduction temperatures and for lower quality feed syngas.

# (600f) Sol-Gel Derived Fe2O3-Based, Al2O3 Supported Oxygen Carriers for Hydrogen Production Via a Chemical Looping Cycle

Yuzbasi, N. S., ETH Zurich
Imtiaz, Q., ETH Zurich
Muller, C. R., ETH Zurich
Originally presented on: 11/7/2013 10:17:00 - 10:39:00

**Sol-gel derived $Fe_2O_3$-based, $Al_2O_3$ supported oxygen carriers for hydrogen production via a chemical looping cycle**

**Nur Sena Yüzbasi, Qasim Imtiaz, Christoph R. Müller***

**Laboratory of Energy Science and Engineering, ETH Zurich, Leonhardstrasse 27, 8092 Zurich, Switzerland**

***Corresponding author: muelchri@ethz.ch**

**Keywords: Hydrogen production, sol-gel, iron oxide, iron precursors.**

Current technologies for hydrogen production such as the steam methane reforming (SMR) process have high production capacities of~15 kte $H_2$/year or greater. However, due to issues associated with the storage and distribution of $H_2$, processes that allow the distributed production of $H_2$ on a small scale may be advantageous.[1] For example, the distributed production of high purity hydrogen from biomass (with simultaneous $CO_2$ capture) can be achieved by modifying the conventional thermo-chemical looping combustion process. Here, biomass is first gasified with steam to produce a synthesis gas i.e. a mixture of predominantly CO, $H_2$, $CO_2$ and $H_2O$. The synthesis gas subsequently reduces iron oxide to a lower oxidation state, thereby producing $CO_2$ and $H_2O$ which can be separated via condensation. Subsequent re-oxidation of the reduced iron oxide with steam yields high purity $H_2$. One of the key challenges of the above described looping process is the synthesis of iron oxide-based oxygen carriers that possess (i) a high reactivity, (ii) stable redox characteristics over many cycles and (iii) tolerance to typical synthesis gas impurities e.g. sulphurous components or tarry substances. Recent studies have demonstrated that pure iron oxide, i.e. unsupported iron oxide, deactivates after only a few cycles when reduced completely to Fe.[1, 2]

A possibility to stabilize the redox characteristics of iron oxide based oxygen carriers is to stabilize $Fe_2O_3$

with a support material e.g. $Al_2O_3$,[3] $ZrO_2$,[4] $MgAl_2O_4$.[5] With regards to synthesis techniques, sol-gel methods have received some attention recently since they ensure a homogeneous mixing of the different components and allow the control of the pore structure and surface area of the material. However, there is only limited work on Fe-based oxygen carriers synthesized via sol-gel techniques.[3, 7] In this study, we prepared different $Fe_2O_3$-based, $Al_2O_3$ supported oxygen carriers (containing 60 wt. % $Fe_2O_3$) to assess critically the influence of the iron precursor, *viz.* iron nitrate ($Fe(NO_3)_3 \cdot 9H_2O$), iron chloride ($FeCl_3$) or iron acetylacetonate ($Fe(C_5H_7O_2)_3$), on the yield of $H_2$. In addition, the effect of the pH on the structure of the synthesized materials was studied. For comparison, pure iron oxide, i.e. unsupported iron oxide manufactured via mechanical mixing technique, was used. The oxygen carriers were characterized using (i) a thermogravimetric analyser (TGA), (ii) a packed bed reactor, (iii) X-ray diffraction (XRD), (iv) $N_2$ adsorption, (v) high resolution scanning electron microscopy (HR-SEM) and (vi) Fourier transform infrared spectroscopy (FTIR). It was found that the oxygen carriers synthesized using iron chloride or iron acetylacetonate as the iron precursor possessed the highest surface areas and pore volumes. In addition, SEM images showed that the use of iron nitrate or iron acetylacetonate as the iron precursor resulted in oxygen carriers that possessed a homogenous and nano-structured morphology. The cyclic $H_2$ production of the oxygen carriers was tested over 15 cycles in a fixed bed reactor. Pure $Fe_2O_3$ possessed a high initial $H_2$ yield of i.e.~ 13 mmol/g oxygen carrier. However, after the first cycle the quantity of $H_2$ produced decreased substantially. On the other hand, $Al_2O_3$-supported oxygen carriers showed stable redox characteristics over 15 cycles. Oxygen carriers synthesized using iron nitrate as the iron precursor showed the highest $H_2$ yield, i.e. ~ 6 mmols/g oxygen carrier which is nonetheless less than the theoretically expected $H_2$ yield. The lower $H_2$ yield is linked to formation of hercynite, $FeO \cdot Al_2O_3$ (as confirmed by the XRD) since its re-oxidation with steam is thermodynamically limited. [7]

**References:**


1. C. D. Bohn, C. R. Müller, J. P. Cleeton, A. N. Hayhurst, J. F. Davidson, S. A. Scott, J. S. Dennis, Ind. Eng. Chem. Res., 47 (2008) 7623-7630.
2. C. D Bohn, J. P. Cleeton, C.R. Müller, S. Y. Chuang, S. A. Scott, J. S. Dennis, Energy Fuels, 24 (2010) 4025-4033.
3. A. M. Kierzkowska et al., Ind. Eng. Chem. Res., 49 (2010) 5383-5391.
4. W. Liu, J. S. Dennis, S.A. Scott, Ind. Eng. Chem. Res., 51 (2012) 16597-16609.
5. J. Adanez et al., Progress in Energy and Combustion Science, 38 (2012) 215-282.
6. F. Li, H. R. Kim, D. Sridhar, F. Wang, L. Zeng, J. Chen, L. S. Fan, Energy and Fuels, 23 (2009) 4182-4189.
7. P. R. Kidambi, J. P. E. Cleeton, S. A. Scott, J. S. Dennis, C. D. Bohn, Energy and Fuels, 26 (2012) 603-617.

**(600g) NiO/Ce-Î³Al2O3 Oxygen Carrier for Chemical-Looping Combustion: Reactivity and Stability**

Mohamed, S.,
Razzak, S., King Fahd University of Petroleum and Minerals
Quddus, M. R., The University of Western Ontario
de Lasa, H. I., University of Western Ontario
Hossain, M. M., King Fahd University of Petroleum & Minerals
Originally presented on: 11/7/2013 10:39:00 - 11:00:00

This study deals with the development of a Ce modified NiO/Ce-$\gamma Al_2O_3$ oxygen carrier suitable for a CLC process. In CLC, the most important characteristics of an oxygen carrier are the reactivity, oxygen carrying capacity and stability under cyclic operation [1, 2]. To investigate these matters successive TPR and TPO experiments were developed in the context of the present study. In cyclic TPR/TPO experiments, the reduction and oxidation profiles of NiO/Ce-$\gamma Al_2O_3$sample remains unchanged and the amount of Ni reduction also shows a stable behavior. Approximately 90 % nickel oxide conversion was achieved using the Ce modified sample, while the value was in the 75 % range for the unmodified carrier. This observation suggests that the addition of Ce helps the formation of easily reducible nickel oxides on the Ce-modified alumina support and minimizes the formation of nickel aluminates (reduces at high temperature only) [3].

The hydrogen pulse chemisorption results (after each TPR cycle) further confirmed the stable behavior of the Ce modified sample. A higher metal dispersion was observed in the Ce modified sample although both the unmodified and modified samples showed a stable metal dispersion with successive redox cycles. This indicates that Ce helps to increase the dispersion of nickel. Thus, the addition of Ce alters the metal surface modifying the degree of interaction between Ni and alumina support, maintaining a consistent metal dispersion during the repeated recox process. It is also apparent that the metal crystal size of the NiO/Ce-$\gamma Al_2O_3$sample remained unchanged over repeated cycles, an indication of the absence of agglomeration of nickel crystals.

**References**

1. M.M. Hossain, H.I. de Lasa, Chem. Eng. Sci., 63 (2008), 4433-4451
2. M.M. Hossain, K.E. Sedor, H.I. de Lasa, Chem. Eng. Sci., 62 (2007) 5464.
3. M. R. Quddus, M.M. Hossain, de Lasa, H.I., Catal. Today (In press).

**(616a) Development of Bench-Scaled Adsorption Type Steam Recovery System for Generating High Temperature Steam From Hot Waste Water**

Kawakami, Y., Takasago Thermal Engineering Co., Ltd.
Abe, Y., ThyssenKrupp Otto
Ito, K., ThyssenKrupp Otto
Marumo, K., Morimatsu Industry Co., Ltd.
Aoyama, T., Morimatsu Industry Co., Ltd.
Tanino, M., Takasago Thermal Engineering Co., Ltd.
Nakaso, K., Kyushu University
Nakagawa, T., Graduate school of Okayama Prefectural University
Itaya, Y., GIfu University
Fukai, J., Kyushu University
Originally presented on: 11/7/2013 8:30:00 - 8:52:00

The efforts for energy conservation are requested because of the limitation of fossil fuel and the environmental issue of global warming. Petrochemical and Steel industries are especially the largest energy-consuming manufacturing industries. While large amount of hot waste water with the range of 60-90 degree C is released from these industrial processes, steam is demanded and large quantities of fossil fuel are generally consumed. The system to generate steam from low-grade waste heat is therefore required for the effective utilization of energy.

In this study, a novel steam generation process using a direct heat exchange system with adsorbent-water pair is proposed. Contacting water with adsorbent makes excess water evaporate due to the release of adsorption heat from adsorbent. Because this system does not require any heat exchangers, increment in packing density of adsorbent particles in the reactor is expected. The purpose of this study is to investigate the novel steam generation system using adsorbent-water pair. Cycle operation which consists of steam generation process and regeneration process of this system has been studied experimentally and numerically using laboratory-scaled equipments of steam generator for 0.3 kg-adsorbent/unit. Based on this basic study, bench-scaled adsorption type steam recovery system of the steam generators for 10-15 kg-adsorbent/unit were manufactured. Operation condition of the steam generation process and the regeneration process of this system was investigated using bench-scaled equipment of one steam generator. And more, continuous steam generation operation was examined using this equipment of the three generator units.

The bench-scaled adsorption type steam recovery system is composed of steam generation system, air regeneration system, and their auxiliaries. The steam generation system is composed of cylindrical steam generators, a feeding pump of water, a vacuum pump, a compressor of air, and a water tank for hot water. The steam generator was made of stainless steel with the height of 565 mm and the inner diameter of 498 mm. As the adsorbent, zeolite pellet was packed into the steam generator for 10-15kg/unit. The air regeneration system produced high temperature dry air for desorbing zeolite pellet at the regeneration process. The air regeneration system is composed of a blower, a desiccant wheel, a pre-heater, and LiBr/H2O absorption heat pump to heat up dry air.

First, operation conditions of the steam generation process and the regeneration process were confirmed experimentally using the bench-scaled equipment of one steam generator. In the steam generation process, air in the steam generator was removed by the vacuum pump. Hot water of 80 degree C was introduced to the steam generator from the water tank by the feeding pump. Steam was then generated by adsorption heat. The steam generation process was terminated when water interface reached the top of packed bed of zeolite pellet. At the last, remained water was drained from the bottom of the steam generator by compressed air from the compressor. After the steam generation process, it is changed to the regeneration process. High temperature dry air at 1050 $Nm^3/h$ was introduced to the generator for 1200 sec. The high temperature dry air with dew point of -40 degree C was produced from ambient air by the desiccant wheel. The high temperature dry air was heated up to 70 degree C by the pre-heater, and more, heated up to 125 degree C by the LiBr/H2O absorption heat pump. As results of those processes, superheated steam whose temperature was 200 degree C and pressure was 0.27 MPa was generated from the hot water at 80 degree C. The peak temperature in the packed bed reached 260 degree C. The mass of steam generated per unit mass of zeolite was 0.09 kg-steam/kg-zeolite/cycle.

After that, Cycle operation condition for continuous steam generation was confirmed using the bench-scaled equipment of the three generator units. Steam was by turns in the three generators (unit A, B, C). The steam generation process time and the regeneration process time were decided to be 600 sec and 1200 sec, respectively. For example, when the unit A was operating at the steam generation process, the unit B and C were operating at the regeneration process. After 600 sec, the unit B was turned to the steam generation process, and then unit C and A were operating at the regeneration process. Furthermore, after 600 sec, the unit C was turned to the steam generation process. By the experimental result, the generated steam whose temperature was 125-155 degree C and pressure was 0.25-0.55 MPa (140 degree C and 0.4MPa on the average) was obtained continuously from the three generator units.

This work was supported by Research and Development Program for Innovative Energy Efficiency Technology under the New Energy and Industrial Technology Development Organization (NEDO) project based on a grant from METI.

**Bench-Scaled Adsorption Type Steam Recovery System**

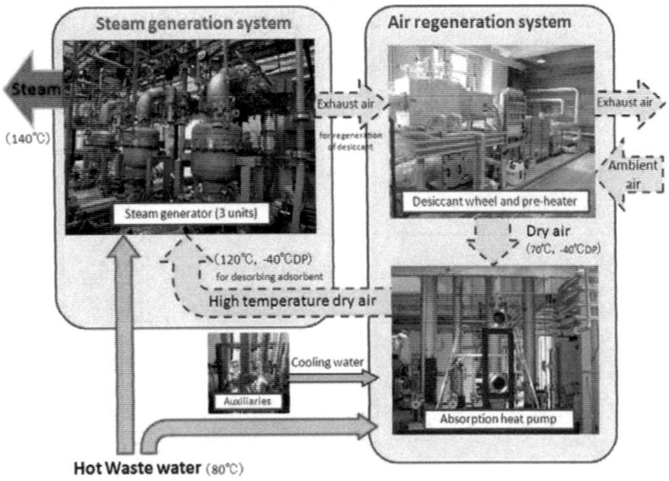

**(616b) Strategic Long-Term Energy Planning for Natural Gas to Liquid Fuels (GTL) Supply Chain Systems**

Elia, J. A., Princeton University
Li, J., Princeton University
Floudas, C. A., Princeton University
Originally presented on: 11/7/2013 8:52:00 - 9:14:00

Interests on natural gas in the United States energy sector have increased in recent years due to the discoveries of shale gas coupled with advancements in horizontal drilling and fracking technologies. The increase in production has lower the cost of natural gas, making it an attractive feedstock for fuels and energy production in the near term. With the abundant natural gas reserves in the country, natural gas is slated to continue as a major energy source for the following decades.

Recently, optimization frameworks for energy supply chain analyses based on hybrid coal, biomass, and natural gas to liquids (CBGTL) and natural gas to liquids (GTL) processes have been proposed [1-3]. The framework includes the locations of the feedstocks in the United States discretized by county, the delivery locations of fuel products, the transportation costs of every input and output of the refinery, the material balances of each refinery, water resources, electricity requirement or production for the supply chain, and the $CO_2$ sequestration capacities in the United States. Formulated as a large-scale mixed-integer linear optimization (MILP) model, the solution gives the strategic locations for the refineries that minimize the total cost of fuel production. The frameworks, however, do not consider the temporal aspect of the problem, namely the different phases of the decisions to implement and build the refineries over a long time horizon.

An optimization-based deterministic strategic planning approach for GTL systems in the United States is proposed in which decisions are made over a 30-year horizon. These decisions include the number of the new refineries built, their locations, capacities, and sources of feedstocks and product flowrates. The 30-year horizon is divided into 10 periods and the decisions are made each period. Multiple capacities of the GTL refineries are considered (i.e., 1, 5, 10, 50, and 200 thousand barrels per day) as well as different fuel ratios produced (i.e., commensurate with the United States demand, maximized diesel production, maximized kerosene production, or unrestricted fuel composition) [4-5]. The model is formulated as a large-scale MILP model that maximizes the net present value of the overall network. Case studies based on states and state groupings are presented.

1. J. A. Elia, R. C. Baliban, X. Xiao, C. A. Floudas. Optimal energy supply network determination and life cycle analysis for hybrid coal, biomass, and natural gas to liquid (CBGTL) plants using carbon-based hydrogen production. *Comp. Chem. Eng.*, 2011:35,1399-1430.

2. J.A. Elia, R.C. Baliban, C.A. Floudas. Nationwide Energy Supply Chain Analysis for Hybrid Feedstock

Processes with Significant $CO_2$ Emissions Reduction. *AIChE J.,* 2012:58, 2142-2154.

3. J. A. Elia, R. C. Baliban, C. A. Floudas. Nationwide, Regional, and Statewide Energy Supply Chain Optimization for Natural Gas to Liquid Transportation Fuel (GTL) Systems. *Ind. Eng. Chem. Res.*, 2013, under review.

4. R.C. Baliban, J.A. Elia, C.A. Floudas. Novel Natural Gas to Liquids Processes: Process Synthesis and Global Optimization Strategies. *AIChE J.*, 2013:59, 505-531.

5. R.C. Baliban, J.A. Elia, V.W. Weekman, C.A. Floudas. Process synthesis of hybrid coal, biomass, and natural gas to liquids via Fischer-Tropsch synthesis, ZSM-5 catalytic conversion, methanol synthesis, methanol-to-gasoline, and methanol-to-olefins/distillate technologies. *Comp. Chem. Eng.*, 2012:47, 29-56.

6. Verderame, P. M., Elia, J. A., Li, J., Floudas, C. A. Planning and scheduling under uncertainty: A review across multiple sectors. *Industrial & Engineering Chemistry Research* 2010:*49*, 3993-4017.

**(616c) Energy Sustainability: Trade-Offs of Consumption, Efficiency, and Environmental Impact**

Vance, L., U.S. Environmental Protection Agency
Eason, T., U.S. Environmental Protection Agency
Cabezas, H., U.S. Environmental Protection Agency
Originally presented on: 11/7/2013 9:14:00 - 9:36:00

It is widely known that one of the major challenges in sustainability is energy. We explore sustainable scenarios and trade-offs building on a previously discussed simple but rigorous model for the relationship between energy production, environmental impact, human population, gross domestic product, and energy efficiency. The theory assumes that: (1) energy production increases gross domestic product and human population, (2) each joule of energy generated has a non-zero impact on the environment, (3) energy production decreases only with improved energy use efficiency, and (4) that the total environmental impact of energy production for 1977 represents an upper bound because in 1977 the human ecological footprint approximately equalled the World biocapacity. Note that here we use the ecological footprint merely as a convenient means of establishing a reference point in time without further implications. Using literature data both actual and projected for GDP, population, energy generation, and energy use efficiency, we develop Pareto curves for the necessary trade-off between environmental impact and efficiency that are necessary to maintain sustainability. Our results indicate that by 2030 the efficiency that would be needed becomes unphysical at more than 100% if the environmental impact of producing a joule of energy remains near that in 1977. However, modest reductions in environmental impact coupled with higher efficiency and lower levels of energy production may offer a path forward. The Pareto curves developed during this project offer a range of choices which will be discussed in detail considering the efficiency and environmental impact of various means of energy production. Further, setting reduction of green house gas emissions as a goal, we have computed the levels of energy production and energy use efficiency that would have to be meet in order that meet the objectives. In summary, the large amounts energy required by modern societies pose a substantial challenge for sustainability, but we believe that with appropriate choices a path to sustainability is feasible.

**(616d) Evaluation of Itev (Air-conditioner Integrated Electric Vehicle) Using in the PV & EV Combined System**

Shibata, S., Graduate school Okayama Prefectural University
Shimono, K., Graduate school Okayama Prefectural University
Nakagawa, T., Graduate school of Okayama Prefectural University
Originally presented on: 11/7/2013 9:36:00 - 9:58:00

After the Great East Japan Earthquake, it has become necessary to reduce nuclear power dependence, as well as $CO_2$ emissions. Smart Community is expected to be one of the solutions. In the Smart Community, energy supply and demand system has been developing to manage with ICT (Information and Communication Technology) to utilize energy efficiently and increase the amount of renewable energy. In one of the studies, the photovoltaic power generation (PV) and electric vehicle (EV) combined system have been proposed. In this system, EV which is utilized as energy storage and transportation can use natural energy for motive power and supplies electricity to a house. So, this two-way system is possible to reduce $CO_2$ emissions drastically with a short payback period. However, it will take a long time before EV becomes widespread, because it's necessary to solve some problems which are high energy consumption of air-conditioner, high cost of storage battery and risk aversion of dead battery. Therefore, air-conditioner Integrated Electric Vehicle (ITEV) which has great possibility to solve the current EV problems has been proposed. ITEV has a hybrid system which activates the air-conditioning system and generates electric power through the use of a micro-engine, a micro-gas turbine or a small SOFC (Solid Oxide Fuel Cell). On the other hand, ITEV which uses fossil fuels may increase $CO_2$emissions compared to EV.

In this paper, the authors have studied the introduction effects of ITEV using in the PV and EV combined system which is considered to optimize not only the vehicle but also as s whole energy system in a Smart community.

As a result, ITEV can reduce $CO_2$ emissions by 37% as the same as current EV in the PV and EV combined system, because ITEV has the hybrid system which generates electric energy in the case of low air-conditioning load. While $CO_2$ emissions of ITEV is increased by using fossil fuels, consequently supplied energy from ITEV to a house is increased and $CO_2$ emissions of the house are reduced. Moreover, ITEV is able to gain the cruising range twice as long as EV with the same battery capacity. ITEV is also able to reduce the cost compared to EV. Because of the long cruising range, ITEV can be utilized as a means of commuting not only short commuting range like Japan but also the long commuting range like Europe and North America. In this paper, the micro-engine efficiency is assumed to be 30%, also heat recovery from the micro-engine is not utilized. ITEV will have a great possibility of further reducing $CO_2$ emissions and improving energy utilization efficiency.

**(616e) Experimental Investigation of Self-Heat Recuperation Using Magnetocaloric Effect for Energy Saving in Thermal Processes**

Kotani, Y., The University of Tokyo
Kansha, Y., The University of Tokyo
Ishizuka, M., The University of Tokyo
Tsutsumi, A., The University of Tokyo
Originally presented on: 11/7/2013 9:58:00 - 10:20:00

The reduction of carbon dioxide ($CO_2$) and the fossil fuel consumption has been one of our major concerns over the past few decades. In chemical processes that involve heating, heat is mainly provided by fossil fuel combustion, which is associated with great exergy loss. In 2009, self-heat recuperation technology based on the exergy recuperative heat utilization principle has been developed. In a self-heat recuperative process, the exergy of process stream heat is recuperated and so that it can be circulated within the process. By compressing gas at its highest temperature in the process for heat elevation, the energy consumption for heating can be reduced drastically compared to conventional heat-recovery technology due to 1) absence of combustion, and 2) small temperature difference during heat exchange.

In the conventional self-heat recuperative process, the recuperation of process stream heat exergy is caused by enforcing a change of state; change of pressure through compression. However, but means to recuperate heat exergy by enforcing a reversible temperature change is not limited to pressure variation. When a magnetocaloric material is subjected to a varying magnetic field, the entropy change of the magnetic moments inside the material enforces a reversible temperature change, which is the magnetocaloric effect (MCE). In this research, an active magnetic regenerative (AMR) heat circulator that utilizes MCE in self-heat recuperation is proposed. Its energy saving potential has been evaluated through a one-dimensional numerical model and its results has been compared with experimental results.

From the simulation, it was found that when the process stream heat of 21.6 J was circulated in between 293 K and 299.5 K, 1.2 J of work is needed. This is equivalent to heat pump with COP 18. The results showed that AMR heat circulator has the potential of being one of the energy saving options for thermal processes where compressors cannot be applied.

**(616f) Energy Diversification in the Middle East: How to Double a Region's Energy Resources**

Kolbe, M., Baker Engineering and Risk Consultants, Inc. (BakerRisk)
Ravaghi-Ardebili, Z., Politecnico di Milano
Originally presented on: 11/7/2013 10:20:00 - 10:42:00

Energy Diversification in the Middle East: How to double a Region's Energy Resources

Massimiliano Kolbe[a], Zohreh Ravaghi-Ardebili*[b]

[a] Baker Engineering and Risk Consultants, Inc.,11011 Richmond Avenue, Suite 700, Houston, Texas

[b] Politecnico di Milano, Dipartimento di Chimica, Materiali e Ingegneria Chimica,?Giulio Natta?, Piazza Leonardo da Vinci 32, 20133, Milano, Italy

*zohreh.ravaghi@mail.polimi.it, mkolbe@bakerrisk.com

Due to the favorable climatic conditions, the Middle East is considered as a one of the best candidates for development and production of renewable (particularly solar) resources. It is estimated that during 80% of the year, these regions are exposed to tremendous quantities of solar radiation. In addition, the majority of these regions are desolate, unusable lands where many of the world's largest oil and gas reserves can be found beneath them. Oil and gas has enriched the Middle Eastern economies for decades and provided a quick path to wealth and development. However, not all Middle Eastern countries have this natural abundance of oil and gas, nor does this resource exist equally everywhere in the regions which do have this abundance. Therefore, considerations must be made in the management of this nonrenewable energy source in tandem with the ever present renewable solar energy source in such a way so as to balance the energy production and consumption. In so doing we can protect, strengthen and grow the economics of the region's oil and gas industries which in many cases are by far the largest source of income for many Middle Eastern nations.

Qatar and Saudi Arabia possess vast fields of natural gas; however it is important to trade and export LNG in order to sustain the economies of these regions. Equally as important, the natural gas can be fractionated to produce NGL components for petrochemical feedstock which are later used in the production of polymers. These countries currently do not fully benefit from other durable sources of income (such as: agricultural, manufactured goods, food processing, etc) in order to balance and sustain the national cash flows. Similarly, Kuwait, Oman and the United Arab Emirates are richer in crude oil resources and cannot benefit as much from petrochemical feedstock from natural gas. Although, Saudi Arabia is plentiful in oil and gas, the produced oil is practically sustaining the economy, and a large portion of the produced gas is driven into the generation of power to supply facilities, buildings and desalination plants. Accordingly, Bahrain and Yemen are not as rich in oil and gas as other countries in this region and are therefore strongly dependent in foreign energy supply for their sustainability.

Fortunately all of these regions are extremely rich in sunlight (and heat). Conversely nearly all of these countries are extremely poor in fresh water resources. Ergo, it is necessary to generate power to not just produce the electricity but to run desalination plants to purify salt water into potable water as well as to run the next highest energy consumption devices; air conditioning and chilling systems. Potable water production from desalination and air conditioning/chilling are vital requirements for such climates and geographies given their need and scarcity. Hence, the produced oil and gas which would practically be the main true source of income is being used as a commodity as well as being used to sustain the power production and infrastructure of the country. Oil and gas are non-renewable and finite, fresh water is already a scarcity in these regions; harnessing solar power and developing this technology as part of the infrastructure to the Middle Eastern region is a vital and logical next step in the quest for sustainability and strengthening the region's economies.

Solar radiation at its peak is above 8000 W-h/m$^2$ in Kuwait alone and can be mentioned as being the largest potential for solar power harnessing in this region and consequently benefiting from common solar technologies for power generation such as photovoltaic and concentrated solar power (CSP) plants taking into consideration the available geographic surface area and the need to generate the power without using the only real sellable commodity; crude oil. The UAE is considered as a second potential for solar radiation availability among the other six countries in the region with an average of 7000 W-h/m$^2$; considering the large modern cities, commercial and industrial facilities harnessing solar power would significantly decrease the national consumption oil and gas which could otherwise be sold internationally. Though the Middle East is considering ramping up the installation of CSP plants this goal will be a long term achievement, on the other hand use photovoltaic panels to capture solar energy for domestic, industrial and commercial buildings is completely tangible in the short term.

Electricity is the appealing and clean form of the power, the efforts to produce it via a technology using renewable and abundant resource would be even more promising. Inasmuch, the idea to produce power for commercial buildings and facilities via photovoltaic solar panels should be strongly considered. Photovoltaic cells in the panels which convert the photons (light) to voltage (electricity); power produced from these panels is in the form of direct current and by means of converters it is then converted to alternating current which is the most applicable form of energy in households and buildings. This procedure of generating the power would be managed individually by the building's owners, industrial administrators and for commercial constructers to establish, collect and provide the power units and their installation along with the tie in to the electrical grid system.

Practically speaking, the path to creating simple yet substantial power sources using photovoltaic solar panels is far more obtainable than once thought due to increased size of cities, industrial facilities and use of automobiles. If we simply considered these public and industrial facilities including the petrochemical sites, malls and hotels typically all hold a large area nearby as parking lots (public car parks) all typically equipped with awnings and shading structures with strong metal frames (see Figure 1 below).

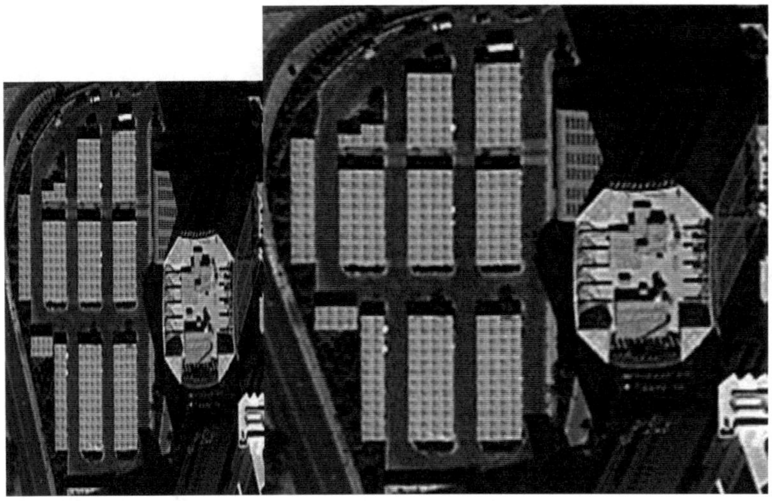

Figure 1. Public building making use of the steel shade parking covers (seen in the red square)

As it can be seen from the typical car park above, these areas are generally very large solid structures made with steel framing and covered with solid material oriented for the purpose of protecting the cars from the sun's radiation. By making use of these strong structures (clearly capable of holding significant loads) and installing photovoltaic solar panels over them, large gathering surfaces can be created. The example above in fact can supply over 1,800 m2 of surface area, with an average solar panel efficiency of 170 W/m2, this particular car park could generate over 300 kW of electrical power. The battery storage or even in some cases grid connections might be easily located in the facility to convert the collected energy and distribute to the nearby building. More importantly in the case of a typical refinery or petrochemical facility, the car parks are much bigger and could potentially produce over 2.5 MW of power with the simple installation of photovoltaic solar panels. Thus, the solar panel collection system would potentially provide all the required lighting or air conditioning for these buildings. It is worth noting that the energy management would not suffer from the logistics of such a system since it is a main part of the energy management.

By prearranged backgrounds, the authors put efforts to propose and plan applicable solutions for the management of energy which could ultimately increase national cash flows going into the country by using photovoltaic solar collecting systems all over cities and in large petrochemical sites throughout Middle East. The discussed and proposed solutions would be provided with detailed calculations for required electrical energy demands, production and consumption, price balances and power generation management to indirectly assist the growth and promote stability in the oil and gas market. Such a movement for using photovoltaic panels or CSP plants could potentially increase the GDP of a typical Middle Eastern country, solve environmental issues and consequently avoid the local use of natural gas to run power plants, petrochemical sites and desalination plants and to divert local high consumption of oil and gas for power generation to the market for sale outside the country and increase cash flows into the country.

In addition, a comparison of energy sources and power generation will be made with the highest energy consumers of the world; USA, China and EU, their energy prices and GDPs and how it relates to their consumption and production of energy. The Middle Eastern countries are generally reliant on oil and gas for their economies and have less consideration of supplementary energy sources. The only approach could be implemented with the integration of energy management which is almost a function of culture and background of this region. Considering the large capital currently produced from the sale of oil and gas from these regions, the available technologies in solar energy generation and the largely available solar radiation in these regions, there is very little risk and high reward in the this solution for increased energy management efficiency.

**(616g) The Total Site Approach As a Synthesis Tool for the Selection of Valorization Paths in Lignocellulosic Biorefineries**

Stefanakis, M. E., National Technical University of Athens
Pyrgakis, K. A., National Technical University of Athens
Mountraki, A. D., National Technical University of Athens
Kokossis, A. C., National Technical University of Athens
Originally presented on: 11/7/2013 10:42:00 - 11:04:00

Since oil reserves are depleted, the production of fuels and chemicals from biomass resulted in the considerable need for the development of biorefineries. In this paper, it is proposed that a Process Integration procedure for under study biorefineries should be combined with a process synthesis tool, since possible chemical production paths constitute important degrees of freedom. The methodology is apparently applicable to any other biorefinery. A Process Integration procedure is significant to secure the feasible and sustainable margins for energy efficiency improvement of chemical processes. Total Site Analysis is a powerful tool of a Process Integration procedure able to consider and manage the complex energy interactions among plants. However, it has been exclusively used to address cases of known and installed plants and not to address cases of processes that are possible to enter in a total biorefinery. Pinch Analysis is also a useful Process Integration tool able to determine the minimum energy consumption and maximum energy recovery of a process.

This paper addresses the use of Total Site Analysis in a process synthesis approach for the valorization and selection of possible chemical paths composing an energy intensive biorefinery. So that, given a scheme of chemical production paths, there are processes that are possible to enter in a final biorefinery. The purpose of this methodology is to select the appropriate combination of processes that subject to minimum energy consumption and offer the highest energy savings using Total Site Analysis upgraded in a process synthesis approach. This purpose gives the guidelines for the development of clusters of processes – each of which represents a Total Site – that are integrated together and include all possible combinations depending on feed distribution and the number of the selected processes each time. Thus, a cluster is able to include as much as possible processes provided that they do not use the same feedstock.

The great advantage of process to process integration is that high available energy of a process is able to be used to cover energy duties of other incoming processes making them very promising. This energy benefit even augments as long as more processes are selected to get integrated together in a cluster. The proposed synthesis approach offers the answers about which processes seem to be the most promising and should be included in the final biorefinery.

The studied Lignocellulosic biorefinery consist of the main process, Organosolv, for the fractionation of biomass to its main components: C5 Sugars, C6 Sugars and Lignin. These intermediate products can be used as available feeds for a range of other chemical processes. Thus, a set of chemical paths, each of which constitutes an individual plant, are branched to convert these intermediates to the final products which are C5 Ethanol and Xylitol (biological and catalytic process) from C5 Sugars, Itaconic Acid and C6 Ethanol from C6 Sugars and PF Resins and PolyUrethanes from Lignin. As a result, there are eight processes that are possible to enter to the final lignocellulosic biorefinery. Energy consumption is targeted in Background Process Heat Exchangers, Evaporator Systems and Distillation Columns. The purpose of this process synthesis approach is to valorize and select the best combination of plants that subject to minimum energy consumption and offer the highest energy benefit.

For the studied chemical production paths, where four feeds are involved (Biomass, C5 Sugars,C6 Sugars and Lignin), there are detected four different types of clusters. Specifically, there are developed Types 1, 2, 3 and 4 including 1, 2, 3 and 4 processes, respectively. As a result, each process can be integrated individually (Type 1) or together with other processes (Types 2, 3 and 4) according to the number of processes selected to be involved in a cluster provided that they use different feedstocks.

Under these circumstances, there are calculated the energy savings of all combinatorial clusters of Types 1, 2, 3 and 4 that may be involved in the final solution. Specifically, there are detected energy savings in ranges of 15 - 87 % for Type 1, 18 - 82 % for Type 2, 22 - 83 % for Type 3 and 67 - 84 % for Type 4. The advantage of integrating more processes together is obviously shown by the increase of the lower limit of the energy benefit ranges proving that existence of available energy is able to be used to cover energy duties for other promising incoming processes. As a result, there are screened and previewed options for the construction of the final lignocellulosic biorefinery.

Therefore, developing this synthesis approach, the selected processes that offer the highest energy savings are Organosolv, BioXylitol, Itaconic Acid and PF Resins integrated in one Total Site (Type 4). The total energy consumption reduced from 226 and 229 MW to 38 and 36 MW of heating and cooling duties, respectively, offering energy savings of 83 % and 84 % for heating and cooling duties, respectively. Additionally, the highest energy savings in another case study where all processes are selected to be included in the final biorefinery resulted in the selection of Organosolv, BioXylitol, Itaconic Acid and PF Resins (Type 4) to be included in the first Total Site, C5 Ethanol, C6 Ethanol and PolyUrethanes (Type 3) to be included in the second Total Site and CatXylitol to be individually integrated (Type 1). Including all processes to the final biorefinery, the total energy consumption reduced from 280 and 282 MW to 79 and 76 MW of heating and cooling duties, respectively, offering energy savings of 72 % and 73 % for heating and cooling duties, respectively.

**(641a) Algal Biorefinery: Production of Biocrude Oil & Byproducts**

Reddy, H. K., New Mexico State University
Selvaratnam, T., New Mexico State University
Muppaneni, T., New Mexico State University
Nirmalakhandan, N., New Mexico State University
Schaub, T., New Mexico State University
Dungen, B., New Mexico State university
Appuhamilage, N. S.,
Lammers, P., New Mexico State University
Voorhies, W., New Mexico State university
Deng, S., New Mexico State University
Originally presented on: 11/7/2013 8:30:00 - 8:52:00

We have developed a strategy needed for the production of sustainable algal biofuels with hydrothermal liquefaction of three different algal species. Hydrothermal liquefaction experiments with 10% solid loading of *Chlorella sp., Nannochloropsis sp.*, or *Galdieria sp.* were carried out between 180°C-300°C and 30 min. of reaction time to investigate the effects of HTL conditions on the important liquefaction reactions and final products. The liquid biocrude products were analyzed with GC-MS and FT-ICR for their chemical compositions, and micro bomb calorimeter for their high heating values. The composition of residual algal biomass was analyzed after each experimental run to identify the biochemical composition of the biomass. The residual water collected after liquefaction was also analyzed to quantify the nutrients and dissolved organic compounds in the water phase. This water phase produced in the HTL process was recycled for cultivation of *galdieria sp.* at different levels of dilution. An integrated processing strategy was developed for individual algal species based on the values of three products (bio-crudes, solid residues and water phase) obtained in the HTL process.

**(641b) Biodiesel Production Through Microwave Assisted Transesterification of Microbial Cells**

Cui, Y., Southern Illinois University

Originally presented on: 11/7/2013 8:52:00 - 9:14:00

One strain of oleaginous yeasts, Cryptococcus curvatus (ATCC 20509) has been studied to grow on several substrates including biodiesel production byproduct crude glycerol and sweet sorghum juice. After cultivation, yeast cells were heated under microwave radiation to extract lipid and produce biodiesel through in-situ transesterification.

Firstly, the yeast growth with crude glycerol was studied.On the second step, an optimization of yeast fermentation with crude glycerol was conducted. The resulting parameters of the response surface method optimization were used in a fed-batch fermentation where crude glycerol was automatically pumped in responding to exhausted oxygen levels in the fermentor.At the end of 12 days, the biomass density and lipid content were 44.53 g/L and 49%. Compared with our previous fed-batch experiment which was conducted under un-optimized condition, the yield of biomass and lipid increased 35.26% and 25.29%.

Considering its capability to utilize a broad range of substrates, C. curvatus was tested on its lipid productivity on sweet sorghum juice. When cultured in a fed batch process for 3 days, the final biomass density was 23.6 g/L with a lipid content of 51%. To extract lipids from cells in an effective and fast fashion, a domestic microwave oven was used with different solvents. With only methanol, a lipid yield of 33.2% of yeast cells was obtained in 4 min.

In the end, to convert yeast lipids to biodiesel directly without the step of lipid extraction,the in-situ transesterification method used microwave irradiation on the simultaneous extractionand transesterification of wet yeast biomass to biodiesel. Based on the experimental results and RSM analysis, the optimal conditions for this process were determined as: methanol to yeast biomass (v:w) ratio of around 50:1, catalyst concentration about 5 wt.%, and reaction time of 2 min. The biodiesel samples were analyzed with GC and the FAME content in biodiesel is about 50%. Compare the with other researches, this process proved to be a fast and easy method to produce biodiesel from wet yeast biomass.

**(641c) Design Optimization By Response Surface Methodology for Continuous Fermentative Production of 1,3 Propanediol From Waste Glycerol By Product of Biodiesel Processes**

Kanjilal, B., University of Connecticut
Noshadi, I., University of Connecticut
Intoci, N., University of Connecticut
Dowding, M.,
Hale, W.,
Brendel, B., University of Connecticut
Srivastava, R., University of Connecticut
Parnas, R., University of Connecticut
Originally presented on: 11/7/2013 9:14:00 - 9:36:00

The production of ~10 gallons of crude glycerol for every 100 gallons of biodiesel has led to a current glycerol excess of ~2.52Bn pounds per annum by the US biodiesel industry and non biodiesel sources. The bio-fermentation of crude glycerol to 1, 3-propanediol, an emerging intermediate platform in biorenewable chemicals with a nascent market of $575 million worldwide, as a value added product gives a recourse to capitalizing on the current over capacity of glycerol. This work presents the design optimization for a green and efficient process for the batch and continuous production of 1,3 propanediol from industrial waste glycerol using soil based bacterial inoculums. The optimal process parameters for the best possible yield of 1,3 propanediol were determined using a design of experiment by varying paramaters such as pH, glycerol concentration and yeast concentration. Response surface methodology (RSM), an important branch of experimental design, was used to find the relationship between experimental parameters and the yield and biomass growth in order to find the optimal conditions for the process. The optimization of 1,3 propandeiol production, in this way, with several inputs, helps to enhance the conversion in an economic manner. pH was seen to have the most profound effect on the process both in terms of yeild as well as bacterial growth. The optimised conditions were used to run a continous fermentation and arrive at a steady state for a given dilution rate, while the inhibitory solvent effects were mitigated by cycling between acetate and butyrate laced batches. While the addition of acetate enhanced biomass growth, butyrate addition enhanced 1,3 propanediol production with distinct inhibitory effects on the growth of bacteria. The effects were utilized to systematically maximize the survival of the mixed inoculum in a continuous run.

**(641d) Review of Bio-Jet Fuel Conversion Pathways**

Wang, W. C., National Renewable Energy Laboratory
Tao, L., National Renewable Energy Laboratory
Originally presented on: 11/7/2013 9:36:00 - 9:58:00

Bio-derived jet fuel has become a key subject in the aviation industry strategy to address the challenges from reducing operating cost, environmental performance and impacts to the global ecosystem. For producing a long-term renewable aviation fuel, economic and environmental sustainability are necessary. Researchers from oil refining industry, commercial aviation industry, government, biofuel companies, agriculture organizations and academic systems are looking forward to developing a commercial viable process that has not only low production cost but low GHG emissions. Additionally, the fuel produced from the process has to meet the ASTM specifications and is intended to be 100% drop-in for the replacement of current petroleum jet fuel. This study reviewed the current major technologies for converting bio-based materials into renewable aviation fuels. All the conversion pathways are classified to four categories: (1) syngas-to-jet, includes Fischer-Tropsch(F-T) synthesis and gas fermentation; (2) renewable-oil-to-jet, includes hydroprocessed renewable jet (HRJ) via plant oil and algal oil, catalytic hydrothermolysis (CH) and hydrotreated depolymerized cellulosic jet (HDCJ, also named pyrolysis) via pyrolysis oil; (3) alcohol-to-jet, includes the conversion of ethanol, butanol and methanol to jet fuel from lignocellulosic biomass feedstocks; (4) sugar-to-jet, includes fermentation of sugars to hydrocarbon or hydrocarbon precursors as well as catalytic conversion of sugars or sugar intermediate to fuels. Discussions for each pathway are primarily focused on the available feedstocks, conceptual process design, process economics, life-cycle assessments and commercial readiness, based on the literature studies. With the help of this comprehensive review on bio-derived jet fuel conversion technologies in recent literatures, it is believed that, whereas there are significant issues related to feedstock price and availability and energy consumption of the process, bio-derived jet fuels could progressively replace a great portion of conventional jet fuel required to meet the commercial and military demand.

**(641e) Pellets From Lignocellulosic Materials: Briquetting Process Evaluation**

Gonzalez, Y., Universidad Libre
Rincon, S., Universidad Libre
Camargo, G., Universidad Libre
Sierra, F., Universidad Nacional
Forero N, C. A., Universidad Nacional

Originally presented on: 11/7/2013 9:58:00 - 10:20:00

Colombia is a palm oil producer country and the implementation of biodiesel industry has generated huge quantities of the residues. On another hand, there are several regions in Colombia without a sufficient energy supply for a better and faster development. The purpose of this research is investigate the use of agro industrial residues coming from palm oil for obtain briquettes that can be used as fuel in remotes region in Colombia. Different parts of the agricultural waste they be employed in the manufacture of briquettes, the fibers and the seed. The biomass was characterized by ultimate analysis and in the densification process two different levels of pressure was employed. Finally densities, mechanical strength, heat of combustion were evaluated for the briquettes. The initial results shown that the waste from the seed materials presents a better performance for the densification process than the fiber materials and the briquettes elaborated a higher pressure exhibit more elevated calorific power

**(641f) Pyrolytic Liquid Yields From Bamboo Species (Bambusa Oldhamii and Bamboo Angustifolia)**

Pliego-Bravo, Y. S., Instituto Tecnologico de Orizaba
Cervantes- Pesado, L., INSTITUTO TECNOLOGICO DE ORIZABA
BolaÃ‘os-Reynoso, E., Instituto TecnolÃ³gico de Orizaba
Gonzalez-Garcia, P.,
Originally presented on: 11/7/2013 10:20:00 - 10:42:00

PYROLYTIC LIQUID YIELDS FROM BAMBOO SPECIES (*Bambusa Oldhamii and Bamboo Angustifolia*)

Abstract

A characterization of the species of bamboo (*Bambusa Oldhamii and Guadua Angustifolia*) with respect to compositional analysis, elemental analysis and thermochemical calorific using as a carrier gas, nitrogen, to identify desirable traits for subsequent use in a thermal conversion process such as pyrolysis, this energetic process provides greater value to convert biomass into a liquid fuel. Moisture content lignin degradation temperatures and ash content are indicators to determine the ideal raw material for conversion to a renewable process.

Introduction

Among the new alternatives that have been implemented for reducing the amount of emissions of CO, SOx, NOx, there being a source of biomass energy that partially replaces fossil fuels and decreases the concentration of these gases emitted into the atmosphere. (Zhang *et al*, 2007). Immediate analysis, elemental and calorific thermochemical analysis are useful in the understanding of the properties of the species to consider a material as an energy source, allowing to know the effective thermochemical process for producing a biofuel (Bridgwater, 2001 and Brown 2011). In this paper we characterize the species of bamboo (GA and BO) to propose an alternative to using craft or promote the efficient use of waste generated by the industry based on the characterizations mentioned above in order to obtain a liquid fuel

Methodology

The proximate analysis, elemental, calorific power and thermogravimetric were performed these are described below.

Materials

In lignocellulosic analysis were used: Ethanol analytical reagent grade (99.5% purity), analytical reagent grade sulfuric acid (98.2% purity), analytical reagent grade toluene, analytical reagent grade nitric acid (69% purity) all of J. T. Baker and distilled water. In thermochemical analysis was used 99% pure nitrogen and air.

Immediate analysis

Establishes the relative amounts of moisture (ASTM D 2867-99), volatile matter (ASTM D 5832-98), fixed carbon, ash (ASTM D 2866-99). Get the amount of carbon fixed subtracting the amounts of ash and volatile matter of a total of 100%.

Elemental analysis and calorific power

The amounts of C, H, N and S of the constituents of bamboo species were estimated experimentally using elemental analyzer (Vario EL III) based on moisture and ash free (LHC), the oxygen obtained by difference. The calorific value was estimated by the equation Reed and Das (1998) expressed in MJ / Kg.

Thermochemical analysis

Thermogravimetric analysis were performed with a non-isothermal model held in the TG Mettler Toledo 851e with software version Star. Samples between 6-10 mg were analyzed, studies were carried out using nitrogen and air, and before each experiment was performed a nitrogen purge for 10 min to all samples with a flow of 100 mL / min.

Getting pyroligneous liquids

For testing proceeded with an experimental $3^2$ where the factors were time (15, 30 and 60 min) and the temperature (350, 500 and 600 ° C) for each of the species studied. These were carried out on a laboratory scale reactor with stainless steel fixed bed capacitors and nitrogen supply 200 mL / min.

Results and discussion

Moistures were between 16-17%, whereby the raw material was suitable for a thermochemical process. The high ash contents cause slag deposits on equipment deteriorating and decreasing the life thereof. The species studied showed percentages of 4.92 of ash for BO and 3.33 for GA making it attractive for use in these applications. Low fixed carbon content was related to the age of the material, in young bamboo lignin was removed easily due to low contents of total polysaccharides and cellulose, whereas an older kind more lignin is harder removing (Garcia *et al*, 2009).

The calorific value was obtained by correlations of Reed and Das (1998). In thermochemical analysis TGA curves showed that the decomposition of the species begins about 300 ° C, the displacement of the curves to the right with increasing temperature. For Bambusa oldhamii degradation stabilized 550 ° C to 600 ° C heating speeds of 15 and 25 ° C / min, while at 5 ° C / min stabilized at 500 ° C. In the case of Guadua angustifolia behavior was observed for more stable speeds of 5 and 15 ° C / min stabilizing at approximately 480 ° C whereas at 25 ° C / min from around 540 ° C. For older BO liquid yields were achieved at 600 ° C with 46 ± 0.5% in the times of 30 and 60 min, for the species GA best performance was obtained at 600 ° C for a time of 60 min of 43 ± 2%.

**Referencias**

[1] Bridgwater, A., V. (2001). *Progress in Thermochemical Biomass Conversion*. Backwell Science. Volume 2 IEA bioenergy.

[2] Brown R. C., *Thermochemical Process of Biomass*. Wiley, U.S.A. (2011).

[3] García, L, Marquez, F, Aguiar, L, Arauso, J, Carballo, L, Orea, U Zanzi, R (2009). *Rendimiento de los productos de la descomposición tÃ©rmica de la madera Eucalyptus saligna Smith a diferentes alturas del fuste comercial.* Revista Chapingo. Serie ciencias forestales y del ambiente, Vol. 15, Núm. 2. Julio_diciembre. pp. 147-154

[4] Higman C. y van der Burgt M., *Gasification*. Segunda edición. Elsevier. U.S.A. (2008).

[5] Reed T. B., y Das A., *Handbook of biomass downdraft gasifier engine system*. Solar Energy Research Institute, U.S. department of energy. U.S.A. (1998).

[6] Zhang Qi, Chang Jie , Wang Tiejun , Xu Ying (2007). Review of biomass pyrolysis oil properties and upgrading research. Energy Conversion and Management 48 (2007) 87–92.

**(641g) The Impact of Ethanol and Iso-Butanol Blends on Regulated Emissions, Air Toxics, and Particle Emissions from SIDI Vehicles**

Karavalakis, G., UC Riverside
Short, D., UC Riverside
Vu, D., UC Riverside
Durbin, T., UC Riverside
Asa-Awuku, A., UC Riverside
Originally presented on: 11/7/2013 10:42:00 - 11:04:00

Spark-ignition gasoline direct injection (SIDI) vehicles have been the subject of several studies, with much of this work has been done outside the U.S., as this is where GDI vehicles have been more prominent. The penetration of SIDI vehicles in the U.S. market is rapidly growing with most domestic automotive manufacturers offering direct injection versions in their existing models. The SIDI technology has emerged from the need to improve the thermodynamic efficiency (and thus reduce fuel consumption) and power output of spark ignited engines. The improved fuel efficiency compared to the conventional port fuel injection (PFI) vehicles could potentially provide the manufacturers the means to meet the target set in the U.S. on the fleet-average carbon dioxide ($CO_2$) emissions of future vehicles. It is foreseen that this category of vehicles will dominate the gasoline market, eventually replacing the conventional and less efficient PFI vehicles. However, SIDI engines are known to emit more particulate matter (PM) emissions than homogeneous-charge PFI spark-ignited gasoline engines and diesel engines equipped with diesel particle filters (DPFs) in the exhaust system. In particular, it is likely that SIDI vehicles will be higher than 3 mg/mile CARB LEVIII PM Standard through 2021, whereas PFI vehicles generally emit below 1 mg/mile, and often below 0.5 mg/mile, providing the potential to more than double PM emissions from new light-duty vehicles. The PM characteristics of direct injected spark ignition engines in comparison to conventional PFI engines can be mainly attributed to the injection characteristics. There is a growing consensus amongst the health experts that particles in the ultrafine range (smaller than 100 nm) are potentially more toxic and have greatest adverse health effects on human health.

Limited studies have investigated the influence of alternative fuels on the exhaust emissions, especially particulate emissions, from SIDI vehicles. The fuels employed in these studies were mainly low and mid-level ethanol blends. Currently, ethanol is the most widely used renewable fuel for transportation in the U.S. and with the push to use increasingly higher levels of renewable fuels, there has been an accompanying push to further increase the ethanol level in gasoline. The EPA Renewable Fuel Standard version 2 (EPA-RFS2) and the California Low Carbon Fuel Standard (CA-LCFS) are driving the U.S. market for the development of biofuels. The EPA-RFS2 requires that 36 billion gallons of renewable fuel are available in the U.S. market by 2022, with ethanol expected to make up the majority of this requirement.

Another pathway that could be used to reach the congressionally mandated biofuel volumes is the use of biobutanol. Analogous to ethanol, butanol can be produced both by petrochemical and fermentative processes. Biomass-derived butanol can be produced by alcoholic fermentation of biomass and agricultural feedstocks, such as corn, wheat, sugar beet, sugarcane, etc. Butanol offers a number of advantages over ethanol for transportation use. Butanol has considerably higher energy content than ethanol. Butanol is less hydroscopic and less corrosive than ethanol, therefore, it is more suitable for distribution through pipelines, whereas ethanol must be transported via rail or truck. Moreover, in blends with gasoline, butanol is less likely to separate from the base fuel than ethanol if the fuel is contaminated with water. Butanol has a lower vapor pressure and a higher flash point than ethanol. The heat of vaporization of butanol is also less than half of that of ethanol.

The literature is particularly poor with respect to studies on butanol/gasoline blends on SIDI engines. In addition the effect of oxygenated fuels especially on particulate emissions from SIDI engines is not well understood. The purpose of this investigation is to elucidate the effects of fuel type and blend concentration for ethanol and iso-butanol on the exhaust emissions from two different modern technology stoichiometric SIDI wall-guided light-duty vehicles fitted with three-way catalysts (TWC). The vehicles were tested over the Federal Test Procedure (FTP) and the Unified Cycle (UC) on light-duty chassis dynamometer. The fuel blends included E10, E15, E20, Bu16, and a E10/Bu8 blend. A total of seven fuels were employed in this study, including an E10 fuel, which served as the baseline fuel, and two more ethanol blends, namely E15 and E20. For this study, iso-butanol isomer was blended with gasoline at proportions of 16, 24, and 32% by volumes, which are the E10, E15, and E20, respectively, equivalent based on the oxygen content. The seventh fuel was an alcohol mixture consisting of 10% ethanol and 8% iso-butanol. This mixed alcohol formulation was equivalent to E15 based on the oxygen content. Each vehicle was tested on each fuel over three FTP and three UC tests. Measurements of regulated emissions were made for nitrogen oxides ($NO_x$), carbon monoxide (CO), total hydrocarbons (THC), nonmethane hydrocarbons (NMHC), methane ($CH_4$), and $CO_2$. Special attention was paid to gaseous air toxics, including carbonyl compounds (aldehydes and ketones) and volatile light aromatic hydrocarbons (benzene, toluene, ethylbenzene, and xylenes), and 1,3-butadiene. Particle emissions were characterized by defining PM mass, total particle number, particle size distributions, and black carbon concentrations. The results of this study will enhance our understanding of alcohol fuel formulations effects on gaseous and particle emissions from emerging vehicle technology.

**(647a) The Novel Cal-C Process for Single-Stage Removal of CO2, SOx and NOx â€“ Experimental Studies and Process Analysis and Integration**

Deshpande, N., The Ohio State University
Phalak, N., The Ohio State University
Wang, W., The Ohio State University
Fan, L. S., The Ohio State University
Originally presented on: 11/7/2013 8:30:00 - 8:55:00

A multiple-gas cleanup process is proposed for the removal of $CO_2$, $SO_x$ and $NO_x$ from coal combustion flue gas. This cleanup occurs at high temperatures (500-650 Â°C) in a single-stage gas-solid fluidized-bed reactor. This process ? named Cal-C process ? utilizes limestone-derived calcium sorbent and inexpensive coal char, and consists of the three-step calcium looping developed at OSU with the addition of char to the carbonator. The calcium sorbent is re-circulated through the system in a closed loop after regeneration and reactivation, while fresh addition of char is warranted due to complete consumption.

The process has been verified experimentally using a lab-scale fluidized-bed reactor setup and the simultaneous removal of all the species of interest has been established. Based on previous studies, lignite coal chars are known to be suitable for this application. In this work, experiments are performed to quantify the consumption of char on the basis of selectivity of char toward NO. Different parameters such as calcium to char loading, $O_2$ concentration and NO concentration are studied for their effect on this selectivity. Further, the fate of gaseous species such as CO and $H_2$, which are likely to be produced in-situ, is investigated and possible process conditions are identified for minimizing unwanted byproducts. Different process configurations with respect to the entry of char in the process loop are also analyzed experimentally. Thermogravimetric analysis of the hydration (reactivation) reaction is also conducted to study the effect of presence of char on the hydration kinetics. Using ASPEN simulations, the process analysis is conducted and different process configurations are evaluated with respect to the opportunities for heat recovery, minimization of excess char requirement, etc.

**(647b) Carbon Dioxide Absorption Into Aqueous Solution of Hydrophilic Amine-Functionalized Ionic Liquids**

Lu, B., Zhejiang University
Wang, X., Zhejiang University
Jin, J.,
Li, W.,
Originally presented on: 11/7/2013 8:55:00 - 9:15:00

Carbon dioxide ($CO_2$), as one of the greenhouse gases (GHG), is currently responsible for over 60% of the enhanced greenhouse effect. Since global warming was aggravated, carbon capture and sequestration to reduce the release of $CO_2$to atmosphere is of critical importance. Chemical solvents (such as amines) have been extensive used, but still suffer some disadvantages, which restricting the promotion of the technology.

In recent years, Ionic liquids (ILs) are frequently used as solvents for gas absorption operations due to their immeasurably low vapor pressure, which can improve the process economy and general efficiency of gas separations. However, gas dissolution in ILs is a physical phenomenon with no chemical reaction and the absorption rate is quite slow. In order to introduce chemical reaction, a variety of ILs were integrated with ions incorporating functional groups. In the open literature, functional ILs were mostly dissolved in conventional ILs and owns high viscosity, which is unbeneficial to industrial applications.

Herein, we designed ether functionalized imidazolium based ILs with glycine anions for the absorption of $CO_2$. And a Hydroxy group was also introduced to this ILs in order to improve the solubility of the ILs in aqueous solution. The absorption of $CO_2$ by hydrophilic amine-functionalized ILs ([$C_2$OHmim]Gly) and Sodium glycinate (SG) were investigated respectively in a double stirred-cell absorber with a planar gas–liquid interface at 15 kPa $CO_2$ partial pressure. The absorption rate and capacity of [$C_2$OHmim]Gly into aqueous solution were higher than SG solution. Moreover, the IR spectra of [$C_2$OHmim]Gly was also proved that the functionalized ILs containing hydroxyl and glycine anions.

It has been widely confirmed that, vaporization and decomposition during regeneration is one of most non-negligible dissipation of aqueous amine, which increases the capital cost of the process. The regenerate efficiency of solvent is important for industrial application. Therefore, the regenerate experiments were investigated in an oil bath at 375.15 ± 0.5 K, and sustained approximately 90 minutes. The multi-cycle regenerate efficiency of [$C_2$OHmim]Gly and SG were compared with each other. The regenerate efficiency were all decreased as the regeneration cycle was increased but [$C_2$OHmim]Gly were decreased quite slow. After the fourth regeneration, the efficiency of [$C_2$OHmim]Gly just decreased 13%, SG almost decreased

25%. The result indirectly proved that the [$C_2$OHmim]Gly were well recycled and has a higher stability than SG by thermal regeneration even after multi-cycle experiments

As oxygen is one of the most typical components in the flue gas and may give rise to undesired side reactions, effect of $O_2$ was investigated and the $O_2$ concentration gradually increased in a range of 0-8 % (V/V). The absorption rate and capacity of $CO_2$ into [$C_2$OHmim]Gly and SG were all decreased in the presence of oxygen, but the decline trend of [$C_2$OHmim]Gly was relatively small. The capacity of [$C_2$OHmim]Gly only reduced 5.9 %, but the capacity of SG almost reduced 17 % in 4%$O_2$ compared with the capacity in absence of $O_2$. Oxygen is susceptible to induce oxidative degradation of alkanolamines. This undesirable breakdown not only reduces their $CO_2$ absorption capacity but also introduces unwanted degradation products, thus forcing the solution to be eventually discarded. Therefore, the multi-cycle regenerate experiments in present of oxygen (4%$O_2$) were also investigated. And the differences of regenerate efficiency of these two solutions were significant.

The above experiment results indicated that the hydrophilic amine-functionalized ILs ([$C_2$OHmim]Gly) we synthesized not only owns a high $CO_2$ absorption rate and capacity but also can enhance the antioxidant capacity of the solution, and can be well recycled after multi-cycle regeneration, possesses great potential for making up the deficiencies of alkanolamine, which is beneficial to industrial applications.

**(647c) The Effect of Tethering Strategies On the Surface Structure of Amine-Functionalized Ionic Liquids: Inspiration On the CO2 Capture**

Xing, H., Zhejiang University
Yan, Y., Zhejiang University
Yang, Q., Key Laboratory of Biomass Chemical Engineering of Ministry of Education, Department of Chemical and Biological Engineering
Bao, Z., Zhejiang University
Su, B., Zhejiang University
Yang, Y., Zhejiang University
Ren, Q., Zhejiang University
Originally presented on: 11/7/2013 9:15:00 - 9:35:00

Amine-functionalized ionic liquids (ILs) are one of the most attractive and promising absorbents for $CO_2$ capture, because of its high performance and notable advantages over traditional organic amine absorbents. For ILs absorbing $CO_2$, it is of primary importance to study the structure and orientation of cations and anions at the interface, which presents a great impact on the surface properties and the transportation of $CO_2$ across the liquid-gas interface. However, despite a variety of cation/anion-tethered amine-functionalized ILs being reported, the choice of the tethering location of the amine group is often somewhat arbitrary and lacks enough consideration in the view of interface chemistry.

Therefore, this work focused on this problem and illustrated the significance of the tethering strategy of the amine functional group on the interfacial structure of ILs for the first time, by investigating the interfacial composition and orientation of cation-/anion-tethered amine-functionalized ILs through molecular dynamic simulation. The results showed that when the amine group is tethered to the cation, strong interaction between the amine group and anions results in a compact interfacial structure without the cation/anion enrichment. However, for amino acid-based ILs with an anion-tethered strategy, the amine group tends to have a higher preference in the outer layer of interface. Therefore, from the viewpoint of interfacial adsorption and transfer, the amine-functionalized ILs with an anion-tethering strategy is considered of great benefit for the $CO_2$ capture, especially when the amine group is on the terminal side of a relatively long alkyl chain.

**(647d) A Carbonate Absorption Process With Crystallization-Enabled High Pressure Stripping for Post-Combustion CO2 Capture: A Kinetic Study and Process Development for Bicarbonate Crystallization**

Ye, Q., University of Illinois at Urbana-Champaign
Lu, Y., University of Illinois at Urbana-Champaign
Originally presented on: 11/7/2013 9:35:00 - 9:55:00

A novel carbonate absorption process with high pressure stripping is currently being developed for post-combustion $CO_2$ capture at the University of Illinois. The process employs an aqueous solution of carbonate, such as $K_2CO_3$, as a solvent. Bicarbonate in the $CO_2$-rich absorber effluent is crystallized by cooling the solution from about 70 °C to 30-35 °C in a crystallizer. The resulting bicarbonate slurry is used for $CO_2$ stripping at a pressure of at least 10 bars. A preliminary techno-economic analysis indicates that the process has an overall energy penalty 30 – 40% lower than that of the bench mark monoethanolamine process, primarily due to lower stripping heat and $CO_2$ compression work requirements.

The phase transition of bicarbonate during cooling of the $CO_2$-rich solution is a key feature of the process. A kinetic study of bicarbonate crystallization has been carried out using a laboratory-scale mixed-suspension, mixed-product-removal (MSMPR) reactor. Parametric tests with respect to agitation rate, mean residence time, temperature, and super-saturation level were performed to evaluate their effects on kinetic rates such as crystal nucleation and growth rates. The results were fit to Mydlarz and Jones (1993) model. Empirical regression formulae were then developed to correlate kinetic rates and crystal dimension with process operating conditions. Based on model predictions, mass and energy balance calculations were conducted to predict the performance of a new design configuration employing multiple continuous stirred tank reactors (CSTR) for crystallizing bicarbonate in the proposed $CO_2$ capture process. This presentation will provide a summary of the kinetic study of $KHCO_3$ crystallization and its implications for the related process design.

**(647e) Carbon Capture and Utilization By Three-Step Aqueous Phase Mineralization**

Hemmati, A., Sharif University of Technoligy
Bu, J., Institute of Chemical and Engineering Sciences (ICES - A*Star)
Shayegan, J.,
Sharratt, P., A*Star. Institute of Chemical and Engineering Sciences
Originally presented on: 11/7/2013 9:55:00 - 10:15:00

The increase in greenhouse gasses concentration in the atmosphere has contributed to global warming in recent decades. Carbon sequestration is one of the solutions to control the $CO_2$ concentration in atmosphere. Â A pH-swing mineral carbonation is the process of extracting magnesium or calcium from mineral rock, followed by precipitation of extracted minerals in separate reactors. The process was found to be feasible method for potential industrial application to reduce $CO_2$ emission by carbon sequestration. This study investigates production of value-added materials for a three-series mineralization reactor in aqueous phase. Magnesium, iron and small portions of some other minerals were extracted from magnesium silicate rock, were precipitated in two steps by increasing the pH from 0.5 up to 10. Product analysis showed that in each step, high purity stable solids such as silica, iron hydroxide II and III, hydrated magnesium carbonate were produced. The $CO_2$ source was $Na_2CO_3$ which was the product generated from the scrubbing reaction between a power plant's flue gas and NaOH to serve as a potential carbon capture method. Mass balance for the mineral sequestration process was investigated for future scale up. $CO_2$ uptake was determined by magnesium carbonate elemental analysis.Â It was calculated that for each ton of sequestrated $CO_2$ approximately 3.22 ton of serpentine rock (mainly $(Mg_3Si_2O_5(OH)_4)$ was required and also 25 $m^3$ of 1M NaOH, 35 $m^3$ of 1M $Na_2CO_3$ and 80 $m^3$ of 1M HCl were utilized. The resultant value-added products from the process would be 3.1 ton of 99.74% pure nesquehonite ($Mg_2CO_3 \cdot 3H_2O$) and 0.82 ton of iron oxide and hydroxide with 90% purity. Crystalline structure, particle size distribution and morphology of hydrated magnesium carbonate produced, were strongly dependent on the last precipitation reactor conditions including temperature, feeding flow rate, mixing speed and base concentration. Additionally, different composition and crystalline structure of magnesium carbonate produced in the last precipitation step in different temperature ranging 0 to 60 Â°C, shows that the process is capable of producing value-added products including nesquehonite, hydromagnesite and dypingite under a wide range of ambient temperature. Â In future investigations, carbon footprint of the process chain will be studied.Â

**(647f) Wood Ash Suspensions in Water As a Solvent in CO2 Absorption Processes**

Petho, S. L., Escola PolitÃ©cnica da Universidade de SÃ£o Paulo
Salvagnini, W. M., Polytechnic School of the University of Sao Paulo
Taqueda, M. E. S., Polytechnic School of the University of Sao Paulo
Originally presented on: 11/7/2013 10:15:00 - 10:35:00

Wood ash is a solid waste with organic and inorganic components, obtained from combustion of wood. Its chemical composition may vary according the tree species, soil type, climate and combustion characteristics. The resulting suspension of ash in water has typical properties of high concentrations of carbonates and high pH values. Together with these qualities, the suspension is an interesting solvent in the study of a $CO_2$ absorption process with low demand of purity and energy.

There are studies about the methodology of determining the calcium carbonate equivalence in ashes because in its composition there other substances like alumino-silicates, silicates and iron oxides that can lead to inaccurate results. So the aim of this study is to develop a simple technique to determine the absorption capacity of unknown ash-aqueous suspensions with carbonates

The experiments were made with wood fuel ashes from a restaurant furnace and validation tests with water and NaOH. The apparatus have a simple characteristics of dismantle all its parts, including the gas tubes and the electrical wirings, forming a called absorption cell, which could be weighted before and after the absorption process.

A simple device was successfully developed that could predict the absorption potential for $CO_2$ of a solvent with unknown composition in ambient conditions of temperature and pressure. The soluble extracted from *Pterodon emarginatus* Vogel. wood, popularly known in Brazil as Faveiro, had 5.33% of calcium carbonate equivalence. The absorption process with an aqueous solution of this solubles, at aprox. 700 mmHg and 20ºC, absorbed 0.77g $CO_2$ / 100g ashes. Ashes from Eucalyptus has 89.74% of calcium carbonate equivalence and its solubles, 11.01%. The absorption process with the ash-water suspension, at aprox. 700 mmHg and 20ºC, absorbed 12.6g $CO_2$ / 100g ashes.

Analyzing the results of pH in function of time suggested that the absorption of $CO_2$ occurred because of the total transformation of the carbonates in bicarbonates. This study suggests that wood fuel ashes are a potential solvent for $CO_2$ absorption.

**(647g) CO2 Chemical Transformation to Valuable Commodities: Toward a Utopia for Fossil-Fuel-Based Economies**

Taheri Najafabadi, A.,

Originally presented on: 11/7/2013 10:35:00 - 11:00:00

Energy and environment have been always classified as the top human challenges by mid-century. In the contemporary fossil-fuel-based civilizations up to 90% of the energy needs are supplied by the unsustainable resources. Therefore, striking a delicate balance between the versatilities of the fossil fuels and their negative environmental impacts (e.g. global warming and climate change) is crucial. Implementing robust strategies to positively impact global carbon balance ultimately leads to the utopia for the fossil-fuel-based economies; with taking full advantage of carbonaceous fuels while minimizing their undesirable effects. This is where carbon capture and sequestration technologies play a critical role to mitigate $CO_2$ from the large scale emitters. Current recipes for the $CO_2$ reduction from large-scale energy consumers (e.g. power stations and cement works) mainly rely on carbon capture and storage (CCS), having three proposed generic solutions: post-combustion capture, pre-combustion capture, and oxy fuel combustion; where the separated carbon dioxide is sent for the physical storage options afterwards. However, researchers believe CCS strategy does not solely resolve the problem permanently, just shifting that from the atmosphere to elsewhere. In fact, long-term ecological impacts of using the earth as a gigantic reservoir for $CO_2$ are off the beaten track. Adding the potential hazards of $CO_2$ leakage to the earth surface noticeably exacerbates the prospect of this approach. Considering all the facts, the ultimate solution of realistically facing with $CO_2$ sequestration concerns is the chemical transformation of $CO_2$ to valuable commodities such as fuels (through, for instance, Fischer–Tropsch chemistry) or polymers (through successive copolymerization and chain growth). This sustainably reduces carbon emissions, remarkably benefiting from $CO_2$-derived chemical commodities, so-called carbon capture and conversion (CCC). Nevertheless, $CO_2$ reduction chemistry is challenging due to the presence of large energy barriers, requiring noticeable catalysis. This work aims to review the most recent advances in the CCC research area selectively (i.e. $CO_2$ conversion to fuels and $CO_2$ copolymerization) from chemical engineering viewpoint in terms of both materials and process design. Some of the most promising studies will be presented in detail with the emphasis on subsidizing more research on $CO_2$ conversion technologies considering the global carbon management urgencies.

**(664a) Evaluation of Spray Dried Oxygen Carriers Based On Manganese Ore and Ca(OH)2 for Chemical-Looping With Oxygen Uncoupling**

Sundqvist, S., Chalmers University of Technology
Mattisson, T., Chalmers University of Technology
Leion, H., Chalmers University of Technology
Lyngfelt, A., Chalmers University of Technology
Originally presented on: 11/7/2013 12:30:00 - 12:55:00

Evaluation of spray dried oxygen carriers based on manganese ore and $Ca(OH)_2$ for chemical-looping with oxygen uncoupling

Sebastian Sundqvist[†,*], Tobias Mattisson[‡], Henrik Leion[†], Anders Lyngfelt[‡].

† Department of Chemical and Biological Engineering, Division of Environmental Inorganic

Chemistry, Chalmers University of Technology, S-412 96 Goteborg, Sweden

‡ Department of Energy and Environment, Division of Energy Technology, Chalmers University of

Technology, S-412 96 Goteborg, Sweden

*Corresponding author

The suitability of an oxygen carrier in chemical-looping combustion (CLC) or chemical-looping with oxygen uncoupling (CLOU) is determined by its lifetime and reactivity. Both are factors relates to the cost of the oxygen carrier in relation to the amount of carbon dioxide captured. There are likely losses of oxygen carrier from attrition in continuous operation and in the case with solid fuels there are additional losses expected together with the ash extraction. Hence there is a need to produce inexpensive oxygen carriers which requires the use of cheap raw materials, such as ores or industrial waste materials.

One type of oxygen carrier that has been shown to have reasonable reactivity is the perovskite structured calcium manganite, or $CaMnO_3$. The oxygen carrier has oxygen uncoupling properties, that is, it releases oxygen in the gas phase, the amount which is governed by the partial pressure of oxygen in the surroundings. In this project, the aim was to produce a $CaMnO_3$ particle from cheap raw materials such as manganese ore and commercial $Ca(OH)_2$ and test these with respect to parameters important for CLC and CLOU.

Several oxygen carriers of CaMnO3 have been produced utilizing different manganese ores as starting material. The manganese ores were mixed with commercial $Ca(OH)_2$ and each material was calcined at both 1200°C and 1275°C for 4h. Particles were then produced using spray-drying and sieved to obtain particles in the size range of 125-180 µm. The evaluation was conducted in a fluidized batch reactor, where the particles were alternatingly exposed to reducing and oxidizing conditions using methane $CH_4$ and syngas (50/50 $H_2$/CO) at 950°C. For the experiment 15 g of oxygen carrier was used with a flow of 450 ml/min of reducing gas. The oxygen carriers were also evaluated at different temperatures (850°C-1050°C) under inert conditions to measure the oxygen release (CLOU properties) and conversion of $CH_4$.

Most of the oxygen carriers have full conversion of syngas but the results for $CH_4$ conversion varies greatly between the different materials, with the average conversion ranging between 40-75% under the reducing period. However the release of oxygen during inert conditions is almost the same for all materials. The oxygen release also increases with temperature, and for some oxygen carriers the oxygen release rate doubles with a temperature increase of just 50°C.

**(664b) Solid Fuel Conversion Via Chemical-Looping With Oxygen Uncoupling (CLOU) Using Co-Precipitation Derived Cu-Mn Oxygen Carriers**

Hosseini, D., ETH Zurich
Imtiaz, Q., ETH Zurich
Kierzkowska, A., ETH Zurich
MÃ¼ller, C. R., ETH Zurich
Originally presented on: 11/7/2013 12:55:00 - 13:20:00

Chemical-looping combustion (CLC) is an emerging $CO_2$ capture and storage (CCS) technology. In CLC oxygen derived from a solid oxygen carrier is used to completely combust a (hydro-) carbonaceous fuel to $CO_2$ and $H_2O$. Thus, in CLC $CO_2$ is inherently separated from the nitrogen in air. Conventional CLC for solid fuels requires a prior, relatively slow gasification step to produce a synthesis gas containing mostly CO and $H_2$. An alternative strategy is chemical-looping with oxygen uncoupling (CLOU) [1]. Here, oxygen is released *via* a decomposition reaction, with the combustion reaction of the solid fuel with molecular oxygen being very fast.

A suitable oxygen carrier for CLOU, typically a transition metal oxide, must possess a high oxygen storage capacity, thermal stability and fast decomposition kinetics at typical operating temperatures (800-1000 ˚C). In addition, resistance to attrition and agglomeration, cost and environmental impact have to be taken into account. With regards to equilibrium thermodynamics CuO and $Mn_2O_3$ are arguably the most attractive oxygen carriers for CLOU [1]. However, both copper and manganese have some drawbacks. For the $CuO/Cu_2O$ system, agglomeration and a comparatively low oxygen partial pressure (at temperatures below 850 ˚C) are the main disadvantages, whereas $Mn_2O_3/Mn_3O_4$ has a low oxygen storage capacity and at temperatures above 900 ˚C $Mn_3O_4$ is thermodynamically stable and cannot be re-oxidized back to $Mn_2O_3$ using air.

In this study, mixed metal oxide oxygen carriers containing manganese and copper oxides were synthesized using a co-precipitation technique. The synthesized oxygen carriers were fully characterized and assessed with regards to solid fuel conversion under CLOU conditions. A comparison with the performance of the individual metal oxides revealed that the optimized Cu-Mn mixed oxides possessed an improved resistance to agglomeration, a higher oxygen carrying capacity, a high solid fuel conversion and an improved physical and chemical stability. These favorable characteristics make the newly developed Cu-Mn mixed oxides promising oxygen carriers for CLOU.

**Reference**

[1] Mattisson, T., Lyngfelt, A., Leion, H., Int. J. Greenhouse Gas Control 2009, 3 (1), 11-19

**(664c) Using (MnxFe1-x)2SiO5 As Oxygen Carriers for Chemical-Looping With Oxygen Uncoupling (CLOU)**

Mattisson, T., Chalmers University of Technology
Jing, D., Chalmers University of Technology
Van Noyen, J.,
Ryden, M., Chalmers University of Technology
Lyngfelt, A., Chalmers University of Technology
Originally presented on: 11/7/2013 13:20:00 - 13:45:00

**Using $(Mn_xFe_{1-x})_2SiO_5$ as oxygen carriers for chemical-looping with oxygen uncoupling (CLOU)**

Chemical-looping with oxygen uncoupling (CLOU) utilizes oxygen carrier materials which can release gas phase oxygen in the fuel reactor. This could be beneficial for the fuel oxidation process, especially when considering solid fuels, as the fuel will be converted through normal combustion, which is not the case with normal chemical-looping combustion. Copper oxide is the most widely investigated oxygen carrier for the CLOU technology, but recently a number of combined oxides have been proposed as feasible oxygen carriers. One especially interesting mixed oxide system is the Fe-Mn-O combination, which releases oxygen at relevant temperatures and partial pressures of oxygen. In earlier investigations by our research group at Chalmers, the behavior and functionality of this system has been shown to be highly dependent upon the molar ratio of Fe/Mn. However, the stability during operation has not been satisfactory, with considerable attrition found when used in continuous operation. Hence it is of interest to combine oxygen carriers of this system with active or inactive material in order to improve stability while keeping the reactivity intact. In this work, the effect of $SiO_2$ addition to the Fe-Mn-O system has been investigated.
Three oxygen carriers based on the Fe-Mn-Si ternary were manufactured by spray-drying and evaluated with respect to parameters important for CLOU. The general composition of the material was $(Mn_xFe_{1-x})_2SiO_5$, with x=0.33, 0.5 and 0.67. After the spray-drying procedure the material was calcined at 1100 and 1200°C for 4h and sieved to a size fraction of 125-180 mm. The oxygen release properties and the reactivity with methane were investigated using a batch fluidized batch reactor made of quartz. The uncoupling properties were investigated in the temperature interval 850 ? 1050°C by exposing a sample of 15 g of particles to a flow of $N_2$, and then measuring the released oxygen as a function of time. For the investigations with methane as the reducing gas, the oxygen carrier particles were exposed to a $CH_4$ flow of 450 Nml/min in the fluidized bed for 20 s. The extent of reaction during the reduction was established by measurement of $CO_2$, CO and $CH_4$ in the outlet from the reactor. After reduction, the reduced particles were oxidized in 5% $O_2$ All of the materials showed oxygen release in the investigated temperature range, although there was not a huge variation with temperature. The $(Mn_{0.67}Fe_{0.33})_2SiO_5$ showed the highest release rate at 900Â°C, with 0.8% $O_2$ measured after 360 s reduction.Â The reactivity with methane increased with temperature for the $(Mn_{0.67}Fe_{0.33})_2SiO_5$ and $(Mn_{0.33}Fe_{0.67})_2SiO_5$ materials, and the average conversion approached 98% for the Mn-rich oxygen carrier. The oxygen carrier with a 1:1 molar ratio of Fe:Mn, i.e. $(Mn_{0.5}Fe_{0.5})_2SiO_5$, had remarkably high reactivity already at 850°C, with an average methane conversion of 90%, which decreased somewhat at 900 and 950°C. The investigated particles had a decent crushing strength, around 1 N, and little or no attrition was seen in the tests. From the current investigation it can be established that the addition of $SiO_2$ to the Fe:Mn system could be feasible, as the oxygen uncoupling properties seem to remain intact and the reactivity with fuel gas high at certain conditions. However, the mechanical stability of these type of materials need to be further investigated, to establish whether there is an advantage in comparison to pure Fe:Mn-oxides.

**(664d) CO2 Utilization Via Chemical Looping Process**

Galvita, V. V., Ghent University
Poelman, H., Ghent University
Marin, G. B., Ghent University
Originally presented on: 11/7/2013 13:45:00 - 14:10:00

**Introduction**

Chemical looping is one of several emerging technology options capable to reduce the carbon dioxide emissions and helping in a diverse range of applications for production of fuels, chemicals and electricity [1-3]. In a general chemical looping process, a metal oxide (MeO) is circulated between two reactors - the reducer and oxidizer. For a combustion process, MeO reacts in the reducer with the fuel to produce $CO_2$, $H_2O$ and leaving metal (Me) or metal oxide in reduced state [1]. In the oxidizer it is re-oxidized to its initial state with air. Thus, fuel and air never mix and $CO_2$ is not diluted by nitrogen.

Other chemical looping schemes can produce hydrogen or synthesis gas. In chemical looping steam (CLSR) or dry reforming (CLDR) processes the fuel is again completely oxidized over the metal oxides (Eq. 1). The $H_2O$ or $CO_2$ product is used as oxidant instead of air. The outlet from the looping reactor then consists of $H_2$ or CO.

A key issue in CLDR processes is the selection of an appropriate oxygen carrier material. The reactivity of the oxygen storage material, its cost and thermal stability are critical selection criteria. A broad range of metals was analyzed to determine the thermodynamic equilibrium limitations for redox cyclic processes [4-7]. While a number of metals/metal oxides (Mo, Cr, Zn, Co, Nb, Ce) give reasonable $CO_2$ reduction capacity, it is apparent that only iron yields high oxygen storage capacity from $CO_2$ (0.7 mol $CO_2$/ mol Fe) over a wide range of operating temperatures (600-1800°C).

Thus, the chemical looping conversion of $CO_2$ into CO or $CO_2$ and $H_2O$ mixture to syngas over iron oxide as oxygen storage material has been one of the possible target technologies for $CO_2$ utilization. The ability of the oxygen storage material to maintain its high activity in repeated reduction/re-oxidation cycles is the most critical issue for the overall economic viability of the $CO_2$ splitting process.

The objective of the present study is to investigate the $CO_2$ re-oxidation efficiency of a series of $CeO_2$-

$Fe_2O_3$ mixed oxides prepared by co-precipitation for the utilization of $CO_2$ in a chemical looping process. Experimental performance data are reported as well as structural characterization of the oxygen storage materials during reduction and $CO_2$ re-oxidation. In addition, the stability of performance was tested through prolonged use of the material.

**Results**

Mixed $CeO_2$-$Fe_2O_3$ materials with 0, 20, 50,70, 90 and 100 wt% $CeO_2$ were prepared for $CO_2$ transformation to CO by means of chemical looping. The samples were characterized with HRTEM, EDX, SAED, EEL and in situ XRD and tested in cycled reactions for activity and stability. The generation of CO from $CO_2$ through $H_2$-$CO_2$ redox cycles was investigated at 600°C to evaluate the effect of $CeO_2$ upon $Fe_2O_3$. Two different phases were observed in the $CeO_2$-$Fe_2O_3$ samples. For a $CeO_2$ content below 70wt%, a distinct $Fe_2O_3$ phase was always present. The second phase appeared in all mixed $CeO_2$-$Fe_2O_3$ samples and was identified as a nano-crystalline solid solution of iron in ceria with 20 nm crystallite size and an atomic ratio of Fe:Ce up to 1:5. In all mixed $CeO_2$-$Fe_2O_3$ samples, the solid solution crystallite size was considerably smaller compared to pure ceria, while a size decrease for $Fe_2O_3$ was noticed with increasing $CeO_2$ loading. The crystallographic structure of all $CeO_2$-$Fe_2O_3$ was studied under $H_2$ reduction and $CO_2$ oxidation, using time-resolved in situ X-ray diffraction. In a 90wt% $CeO_2$ sample, no iron related diffractions were present at the start, but the $H_2$-TPR treatment made Fe appear as phase segregating from ceria. For lower $CeO_2$ loadings, $H_2$-TPR reduced the separate $Fe_2O_3$ phase to $Fe_3O_4$, FeO and Fe. The solid solution phase was always partially reduced. In $CO_2$-TPO, re-oxidation of iron followed a one-step pathway to $Fe_3O_4$ around 500°C, while for the solid solution re-oxidation occurred at 450°C. The addition of $CeO_2$ to $Fe_2O_3$ had a beneficial effect upon the activity and stability of the material. Reduction-re-oxidation cycles could be repeated many times without significant loss of $CO_2$ conversion efficiency.

The activity and stability of the mixed $CeO_2$-$Fe_2O_3$ samples as compared to pure $Fe_2O_3$ and $CeO_2$ a test reaction with $H_2$ reduction and $CO_2$ re-oxidation was set up. The samples are reduced in 5%$H_2$ in Ar during 1 min. By contacting the reduced sample with $CO_2$, CO is produced. As an example, the observed space time yield of CO as a function of time for 20wt% $CeO_2$- $Fe_2O_3$ during the 2nd, 5th and 10th cycle is displayed in Fig. 1a.

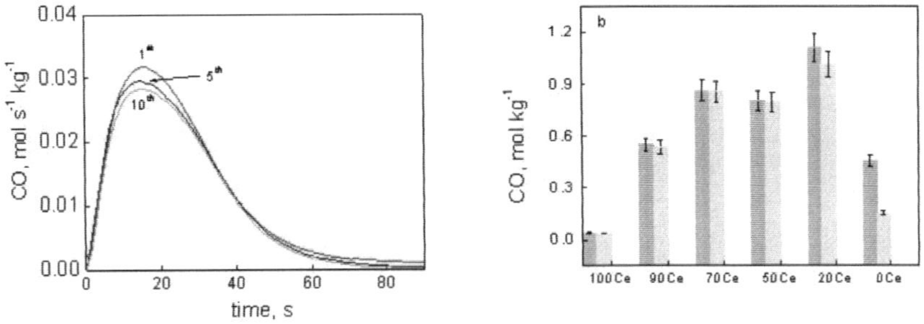


Figure 1. (a) Space time yield of CO during the re-oxidation phase with $CO_2$/He for 20wt% $CeO_2$-$Fe_2O_3$ after 1, 5 and 10 redox cycles; (b) CO produced after $1^{st}$ (dark bar) and $10^{th}$ cycle (light bar) over the xCe-Fe samples.

Fig. 1b shows the amount of produced CO after the $1^{st}$ and $10^{th}$ redox cycle for samples 0 to 100wt%$CeO_2$-$Fe_2O_3$. In both cycles, the product yield is higher for mixed materials than for pure ones. Especially the samples 20 to 70wt% $CeO_2$-$Fe_2O_3$ produce over 5 times more CO than $Fe_2O_3$, with the highest yield for 20 wt%$CeO_2$-$Fe_2O_3$. The amount of CO produced over $CeO_2$ was 10 times lower than for $Fe_2O_3$. Highest CO yield was obtained for 20wt% $CeO_2$-$Fe_2O_3$, while 70 and 50wt% $CeO_2$-$Fe_2O_3$ were the most stable combinations. The addition of ceria to $Fe_2O_3$ thus formed a solid solution phase, yielding improved activity.

Moreover, the combination of both oxides efficiently suppressed the sintering effect, resulting in improved stability in chemical looping processes.

**Acknowledgment**

This work was supported by the 'Long Term Structural Methusalem Funding by the Flemish Government', the Fund for Scientific Research Flanders (FWO, project 3G004613), the Interuniversity Attraction Poles Programme, IAP7/5,- Belgian State – Belgian Science Policy.

**References**

1. J. Adanez, A. Abad, F. Garcia-Labiano, P. Gayan, and L.F. de Diego, Progress in Energy and Combustion Science 38 (2012) 215

2. L.-S. Fan, and F. Li, Ind. Eng. Chem. Res. 49 (2010) 10200.

3. C. Luo, Y. Zheng, N. Ding, Q. Wu, G. Bian, and C. Zheng, Ind. Eng. Chem. Res. 49 (2010) 11778.

4. M. Najera, R. Solunke, T. Gardner, and G. Veser, Chemical Engineering Research and Design 89 (2011) 1533.

5. A. Sim, N.W. Cant, and D.L. Trimm, Int. J. Hydrog. Energy 35 (2010) 8953-8961.

6. S. Abanades, Int. J. Hydrog. Energy 37 (2012) 8223.

7. J.R. Scheffe, M.D. Allendorf, E.N. Coker, B.W. Jacobs, A.H. McDaniel, and A.W. Weimer, Chemistry of Materials 23 (2011) 2030.

**(664e) The Synthesis of Highly Efficient Ca-Based CO2 Sorbents Containing a Hierarchical Pore Structure**

Broda, M., ETH Zurich
Muller, C. R., ETH Zurich
Originally presented on: 11/7/2013 14:10:00 - 14:35:00

In 2010 the global $CO_2$ emissions increased by more than 5 % compared to 2009, reaching 33 billion tonnes.[1] One possible, mid-term solution to reduce the release of $CO_2$ into the atmosphere is the implementation of $CO_2$ capture and storage (CCS) technologies.[2] However, the currently available $CO_2$ capture technology, i.e. amine scrubbing, is associated with a large penalty on plant efficiency. A promising, new $CO_2$ capture technology is the use of alkaline earth metal based $CO_2$ sorbents. Due to its (i) high theoretical $CO_2$ capture capacity of 0.78 g $CO_2$/ g sorbent, (ii) low cost and (iii) relatively vast abundance in naturally occurring minerals, e.g. limestone or dolomite, CaO is the most promising candidate in this class of materials. The selective capture and release of $CO_2$ is achieved by the following, theoretically reversible, reaction:

$$CaO(s) + CO_2(g) \ll CaCO_3\ (s),\ DH^{0298K} = \pm\ 178\ kJ/\ mol \quad (1)$$

However, CaO produced via the calcination of naturally occurring materials has a serious drawback, viz. the rapid decrease in the $CO_2$ capture capacity with repeated cycles of carbonation and calcination.[3] The carbonation reaction can be divided into two regimes.[4] It has been argued that the fast, kinetically controlled reaction stage, is associated with the filling of small pores. The importance of the available pore volume can be clearly seen by comparing the molar volume of the product, $CaCO_3$ ($M_V$ = 36.9 $cm^3$/mol), with that of CaO ($M_V$ = 16.7 $cm^3$/mol). Once the pores have been filled, the second, diffusion controlled reaction stage takes over, in which $CaCO_3$ is deposited on the partially carbonated CaO grains. Owing to the typically short contacting times in circulating fluidized beds, the reactor of choice for the commercial implementation of the calcium looping processes, only the $CO_2$ uptake achieved during the fast reaction stage will be of practical relevance. Since the Tammann temperature (temperature at which thermal sintering starts) of $CaCO_3$ (533 °C) is relatively low compared to the operating temperatures of 650-950 °C, the dramatic reduction in $CO_2$ uptake has been attributed to the reduction of surface area and the blockage of pores with diameter $d_{pore}$< 100 nm. One potential strategy to increase the $CO_2$ capture stability of Ca-based sorbents is to stabilize CaO with high Tammann temperature matrix. However, work in this field has largely focused on the use of ?simple? synthesis techniques, which do not allow easily to adjust key structural properties, such as the pore volume. In this work we report the preparation of carbon gel templated, Ca-based $CO_2$ sorbents. Here, calcium nitrate and aluminium nitrate were used as the calcium and aluminium precursors, respectively, whereas resorcinol and formaldehyde were the precursors of the carbon gel. In addition, the diameter of the carbon sphere template was varied by changing the surfactant (CTAB) to resorcinol ratio. Upon removal of the carbon gel via calcination in air, hollow micro-spheres were obtained. The synthetic material was investigated using (i) scanning electron microscopy (SEM), (ii) surface area (BET) and pore volume (BJH) measurements, (iii) X-ray diffraction (XRD) measurements and (iv) mercury porosimetry measurements. The material synthesized possessed an excellent $CO_2$ capture capacity when compared to limestone or alternative $CO_2$ sorbents, such as hydrotalcites or metal organic frameworks. Furthermore, the current synthesis techniques allow the amount of $Al_2O_3$ required to stabilize the cyclic $CO_2$ capture capacity of the material to be reduced to only 5 wt.%.

## References

<>1. Olivier, J. G. J.; Janssens-Maenhout, G.; Peters, J. A. H. W.; Wilson, J. *Long-term trend in global $CO_2$ emissions. 2011 report;* PBL Netherlands Environmental Assessment Agency: The Hague, **2011**

<>2. Herzog, H. What future for carbon capture and sequestration? *Environ. Sci. Technol.* **2001**, *35*, 148-153.

<>3. Grasa, G. S.; Abanades, J. C. $CO_2$ Capture Capacity of CaO in long series of carbonation/calcination cycles. *Ind. Eng. Chem. Res.* **2006**, *45*, 8846-8851.

<>4. Dennis, J. S.; Pacciani, R. The rate and extent of uptake of $CO_2$ by a synthetic, CaO-containing sorbent. *Chem. Eng. Sci.* **2009**, *64*, 2147-2157.

**(664f) Li4SiO4-Based Novel Solid Fluidizable Sorbent for CO2 Capture:**

Chowdhury, M. B. I., The University of Western Ontario
de Lasa, H., University of Western Ontario
Quddus, M. R., The University of Western Ontario
Originally presented on: 11/7/2013 14:35:00 - 15:00:00

This study investigates the cyclic use of several sorbents for $CO_2$ capture and comparative study of thermodynamic analysis to predict sorbent regeneration conditions and thermal levels where sorbent kinetics change from $CO_2$ absorption to $CO_2$ desorption. It is expected cycling processes with $CO_2$ capture and sorbent regeneration and alternative sorbents (e.g. lithium ortho-silicate based novel sorbent) could lead to significant energy savings with limited sorbent sintering. On this basis $CO_2$ absorption and desorption were quantified at various temperatures and $CO_2$ partial pressures. Temperature Programmed Carbonation (TPC) and Temperature Programmed Decarbonation (TPDC) were developed using temperature programmed equipment at CREC-UWO laboratories employing several sorbents subjected to a 5°C/min temperature ramp and with a gas stream containing a 10% $CO_2$ mole fraction. Calcium carbonate, lithium orthosilicate and a novel lithium orthosilicate modified sorbent are considered for these runs. TPC-TPDC runs confirm thermodynamic predictions for thermal inversion points. Furthermore, TPD-TPDC runs show that the novel lithium orthosilicate based sorbent provides a very stable and increased $CO_2$ sorption capacity over 10 absorption-regeneration cycles, while calcium carbonate displays a reduced $CO_2$ sorption capacity with cyclic operation. This fluidizable novel sorbent can significantly contribute towards $CO_2$ removal from flue gases emitted by power plants. The thermodynamic and TPC-TPDC analysis of the present study also provides a conceptual design from developing $CO_2$ capturing processes with alternative sorbents such as lithium ortho-silicate based sorbent.

## (703a) Isolation, Identification and Characterization of High CO2 Tolerant Microalgae for Efficient CO2 Sequestration From Two Different Locations of India

Ghoshal, A. K., Indian Institute of Technology Guwahati

Originally presented on: 11/7/2013 12:30:00 - 12:55:00

Isolation, Identification and Characterization of high $CO_2$ tolerant microalgae for efficient $CO_2$ sequestration from two different locations of India.

Samarpita Basu[1], AbhijitSarma Roy[2], K.Mohanty[1, 3], A.K.Ghoshal*[1, 2, 3]

[1]Centre for Environment, Indian Institute of Technology Guwahati, Assam-781039, India

[2]Centre for Energy, Indian Institute of Technology Guwahati, Assam-781039, India

[3]Department of Chemical Engineering, Indian Institute of Technology Guwahati, Assam-781039, India

*Corresponding author (E-mail:aloke@iitg.ernet.in)

The accumulation of greenhouse gases like $CO_2$ in the earth's atmosphere as a result of human activities and industrialization causes serious consequences like global warming thereby increasing the temperatures of the surface air and subsurface ocean. Apart from other causes combustion of fossil fuel for energy is one of the most important for increased $CO_2$in the atmosphere. Emissions of $CO_2$ are likely to increase significantly in the near future due to increased energy demand from developing countries.Public concerns over severe environmental consequences that might result due to globally rising temperatures have brought forward the necessities to find out ways to reduce the emission of $CO_2$ to the atmosphere. Some of the strategies employed to reduce $CO_2$ emissions include the implementation of renewable energy sources as well as capturing of $CO_2$ to limit its release into the atmosphere. The current methods of capturing $CO_2$ include the use of physicochemical absorbents and mineral carbonation of $CO_2$.However, these methods have several drawbacks like high space requirements, high cost and potential $CO_2$ leakage over time. Therefore, a lot of attention is now being given to the biological fixation of $CO_2$. Photosynthetic microorganisms like microalgae and cyanobacteria use inorganic carbon for growth and hence can convert $CO_2$ from a point source into biomass. Among the photosynthetic microorganisms, microalgal capture of $CO_2$ is thought to be a feasible technology with energy efficient and environment friendly natureMostly microalgae are unicellular photosynthetic microorganisms that can fix $CO_2$ effectively from gaseous effluents (e.g., industrial flue gas) and have much higher photosynthetic efficiency and biomass productivity as compared to other energy crops Moreover, the microalgae biomass produced after $CO_2$ fixation can be put to several uses such as feed stock for a variety of biofuels, medications, cosmetics and nutritious foods, representing additional benefits from the microalgal $CO_2$ reduction process. The present study aimed to isolate high $CO_2$ and temperature tolerant microalgae capable of sequestering $CO_2$ from flue gas. Water samples were collected from two different locations, one from a freshwater body (26°11'15"N 91°45'4"E) of Guwahati in Assam, located in north eastern region of India and another from a freshwater body (22°55′22″N88°22′46″E) located in the vicinity of a thermal power plant in West Bengal, located in eastern region of India. Isolation of microalgae strains were carried out from the samples using BG 11(ATCC Medium 616) as the culture medium *.14 hours of light and 10 hours of dark regime was employed during growth using daylight-type 20 W fluorescent tubes and light intensity maintained was 5496 lux. In totality, three microalgal strains designated as SA1 (from Assam), SA2 and SA3 (from West Bengal) were isolated. After isolation of pure culture, $CO_2$ fixation studies were initiated using these strains using* 13.8 ± 1.5 % *$CO_2$ as the sole carbon source for microalgal cultivation.* 28s rRNA gene sequencing of SA1 followed by homology matching with NCBI GenBank database confirmed SA1 to be the green microalga, *Scenedesmus obliquus*. The 28s rRNA gene sequence was then submitted to NCBI Genbank and accession number (KC733762) was obtained. Growth kinetics and $CO_2$ fixation rates of the isolates were determined at varying culture conditions (nitrate and phosphate concentrations, photoperiods and temperatures). Studies were carried out using these strains at elevated cultivation temperatures of upto 40°C since flue gas has a temperature of around 40°C after the desulfurization process. Characterizations of the strains were done in scanning electron microscope as well as FTIR. FTIR gave an idea of the biochemical composition of the microalgal strains. Carbohydrate, protein, lipid and pigment content of the microalgal strains were determined to elucidate the potential of these strains for the production of value-added substances like biofuels which would offer additional benefits to the $CO_2$ sequestration process. The various growth kinetic parameters like maximum biomass value, maximum specific growth rate, overall biomass productivity and $CO_2$ fixation rates were estimated for the strain SA1.The microalgae strains SA2and SA3 isolated from West Bengal are currently being studied and DNA isolation and sequencing are being carried out. The gross calorific values of the strains were estimated to elucidate the lipid content of the microalgae strains and also to validate the use of the algal biomass for biofuel generation after the carbon sequestration process. SA1 showed enhanced $CO_2$ fixation rates and high biomass productivities which were higher than most of the related studies. At 13.8 ± 1.5%$CO_2$ and *25 °C,* maximum biomass (4.975 ± 0.003 $gL^{-1}$) and maximum $CO_2$ fixation rate (252.883 ± 0.361$mgL^{-1}d^{-1}$) were obtained. *At* elevated *temperature (40°C) and* 13.8 ± 1.5% *$CO_2$, maximum biomass(*0.883 ± 0.001 g $L^{-1}$) was obtained. *The* carbohydrate, protein, lipid, and chlorophyll content of the SA1 after $CO_2$ sequestration were 30.87 ± 0.64%, 9.48 ± 1.65%, 33.04 ± 0.46 and 6.03 ± 0.19% respectively, which were higher than previous reports. The Gross calorific value (GCV) was 23.02 MJ/Kg for SA1 after $CO_2$ sequestration which was significantly higher than most of the previous reports.

Keywords: Global warming, microalgae, $CO_2$ fixation, flue gas, *Scenedesmus obliquus*

**(703b) Complex Energy System Which Combined Photovoltaic Power Generation With Electric Industrial Vehicle-Case Study of Maniwa SMART Factory-**

Shimizu, A., Okayama Prefectural University
Notoji, Y., Graduate school of Okayama Prefectural University
Nakagawa, T., Graduate school of Okayama Prefectural University
Originally presented on: 11/7/2013 12:55:00 - 13:20:00

SMART factory project has been developed under a grant from the Japanese government at Maniwa biomass base in Okayama Prefecture. In this developmental study, the following two research objects have been addressed. One is to build a new technology concerning manufacturing of nanofibers with high additional value from wood thinned from forests and unused wood parts from sawmills, and the other is to develop a new energy system and the elemental technologies which are provided renewable energy to meet the energy demand at the new factory producing nanofibers. The research target is to reduce economically $CO_2$ emissions of the entire factory by introducing these technologies.

A present factory which is built at Maniwa base has been producing wood chips from forest thinnings and two kinds of energies as light oil and electric power are consumed. Electric power has been used as the power of chip manufacturing equipment mainly. Light oil has been consumed as fuel for forklift trucks, shovel loaders and others. Electric power consumption is caused by 60% of $CO_2$ emissions from the factory and the remaining $CO_2$ emissions are due to light oil. Therefore, only the reduction usage of electric power is not enough to reduce significantly the $CO_2$ emissions of the factory. So, it is necessary to reduce the total consumption of electric power and light oil. In previous study, as a way to reduce electric power and fuels of vehicle simultaneously, the authors have been proposed a method according combined system of photovoltaic power generation (PV) and electric vehicle (EV).

In this paper, the new energy system has studied that industrial vehicle such as forklift trucks become exchanged to drive electricity which is supplied by PV, and PV power fluctuation become leveled using the storage battery in the industrial vehicle.

As a result, it is able to reduce $CO_2$ emissions by 25% according to introduce the new energy system that forklift trucks are driven by PV power which is charged directly to the battery in forklift trucks and is supplied to the factory if the surplus electricity is generated. In addition, it has a great possibility to reduce $CO_2$ emissions by 50% according to introduce an advanced system that shovel loaders and grapples will be driven by PV as the same as forklift trucks.

**(703c) Multi-Objective Optimization of Bioelectricity Supply Chain With Life Cycle Assessment and Social Impact Analysis**

Slivinsky, M., Northwestern University
Yue, D., Northwestern University
You, F., Northwestern University
Originally presented on: 11/7/2013 13:20:00 - 13:45:00

## Multi-Objective Optimization of Bioelectricity Supply Chain with Life Cycle Assessment and Social Impact Analysis

Maxim Slivinsky, Dajun Yue, Fengqi You*

Submitted to session 23C01 Sustainable Electricity: Generation and Storage

### Abstract

There is a strong drive for exploration of bioelectricity production, which is rooted in the increasing costs and limited availability of fossil fuels as well as the desire for greater energy security. Bioelectricity provides a promising alternative to fossil fuel power generation due to the compatibility with power generation infrastructure, reduction in greenhouse gas emissions, and associated job creation. The existing power infrastructure, including power grid, steam turbines, etc., can be utilized in the production of bioelectricity given some equipment modifications at the plant level and in pre-processing facilities [1]. This results in a relatively low capital investment opportunity to produce electricity from biomass feedstocks. This work approaches the supply-chain problem using three objectives: economic, environmental, and social. The economic objective is minimization of total annualized costs, the environmental objective seeks to minimize the life cycle greenhouse gas emissions (GHG), and the economic objective is to maximize total accrued jobs.

In this work a multi-objective mixed-integer linear programming problem (MILP) is developed to model the supply-chain optimization of bioelectricity production from feedstocks such as corn stover, forest residues, and switchgrass. This MILP is solved using the ε-constraint method, where the Pareto-optimal solutions are determined by solving MILP sub-problems with varying values of ε [2]. A key constraint in the formulation is the minimum fraction of annual electricity produced from biomass for a given region. The MILP is analyzed through different scenarios corresponding to this constraint parameter. Economic multipliers are taken from the Jobs and Economic Development Impact (JEDI) model to determine the social impact on the region [3]. The total accrued employment consists of direct employment (e.g. construction), indirect employment (jobs in upstream supply chain), and induced employment (jobs from money spent locally). Since JEDI has data for only two boiler technologies, this formulation looks at circulating fluidized bed and stoker boilers. The economic objective takes into account feedstock and utilities costs, transportation costs, construction costs, electricity demand by region, and processing limits. The environmental objective uses life cycle analysis (LCA), considering gate-to-gate environmental impact analysis of the preprocessing facilities, transportation, and capacity expansions. GWP is calculated using the 100 year timeframe per the Kyoto Protocol.

This problem is evaluated through a county-level case study for Illinois, which consists of 102 counties. County data is used to determine geographic biomass availability, existing power plant locations, and electricity demand. The problem is solved to determine the optimal locations and capacities of preprocessing facilities and biomass technology expansions, as well as the network design and technologies used [4, 5]. The Pareto solutions to this MILP reveal the tradeoffs among the decision variables under economic, environmental, and social objectives. The scenario analysis shows the economies of scale associated with replacing a fraction of power generation with bioelectricity.

### References

[1] Environmental Protection Agency [EPA], Combined Heat and Power Partnership (2007, September). Biomass Combined Heat and Power Catalog of Technologies, http://www.epa.gov/chp/documents/biomass_chp_catalog.pdf

[2]. H. Gebreslassie, Y. Yao, and F. You, "Design under uncertainty of hydrocarbon

**(703d) Techno-Economic Modeling and Analysis of Energy Storage Options for a Self-Sustained Solar PV Array**

Peterson, B. A.,
Smith, J. D., Idaho National Laboratory
Originally presented on: 11/7/2013 13:45:00 - 14:10:00

Renewable energies, such as solar and wind, are an ever-growing part of energy consumption world-wide. The primary push to grow renewable energy production is two-fold: reducing carbon dioxide emissions, thus moderating the effects of increased atmospheric levels of the gas, and reducing dependence on a diminishing fossil fuel supply. However, renewable energies certainly have issues, chiefly their intermittent and non-dispatchable nature. This is to say that solar power is only available when the sun shines, and wind is only available when the wind blows. This makes supplying electricity to a given demand load impossible for stand-alone renewable energy sources, limiting the financial incentive to invest in renewable energies. Intermittency also causes frequency and voltage instability for the grid when renewables are integrated into it. Energy storage is integral to the viability of renewable energy as it allows for load-smoothing and frequency regulation. Currently, no one or two obvious energy storage technologies exist to economically meet the requirements of meeting demand with renewable energy.

There are numerous ways in which energy can be stored: chemically, electrochemically, mechanically, and electrostatically. A model is proposed to study the economic feasibility of several energy storage options available today for a self-sustained solar PV array. Hourly demand data and hourly solar generation were used to determine constraints on the studied energy storage component. Factors including system capital cost, system operating cost, various replacement cost, and energy storage replacement times were used to calculate a net present value of the energy storage options for different demand curves. The model was used to examine the full lifetime of the PV array, typically 25 years. The most promising energy storage technologies for financial feasibility of a generalized solar PV array are sulfur-sodium batteries and carbon-enhanced lead acid batteries, depending on the scale, timeframe, and location of the energy storage.

**(703e) Solar Electricity Generation System With Combined Energy Storage and Steam Generation**

Xu, R., Texas Tech University
Wiesner, T., Texas Tech University
Originally presented on: 11/7/2013 14:10:00 - 14:35:00

Solar steam power plants for electricity generation were launched commercially in 1984. These plants are known as SEGs projects. The oil as the heat transfer fluid (HTF) was pumped through solar field to capture solar energy. However, the oil was unable to maintain the temperature during off-sun periods. Thermal energy storage (TES) is useful to increase the solar multiple. Molten salt is usually employed as a stable energy storage medium. Two-tank and single-tank construction have been employed for parabolic trough-based solar steam power plants. In our paper, we propose an improved thermal storage concept aiming at increasing the operation hours and reducing construction cost. Our design is shown in the figure below(Figure 1,right), which incorporates the steam evaporator inside the single thermal storage tank. Aspen HYSYS is used to dynamically simulate the process, analyze the steam generation efficiency and compare with the single-tank base case(Figure 1, left).

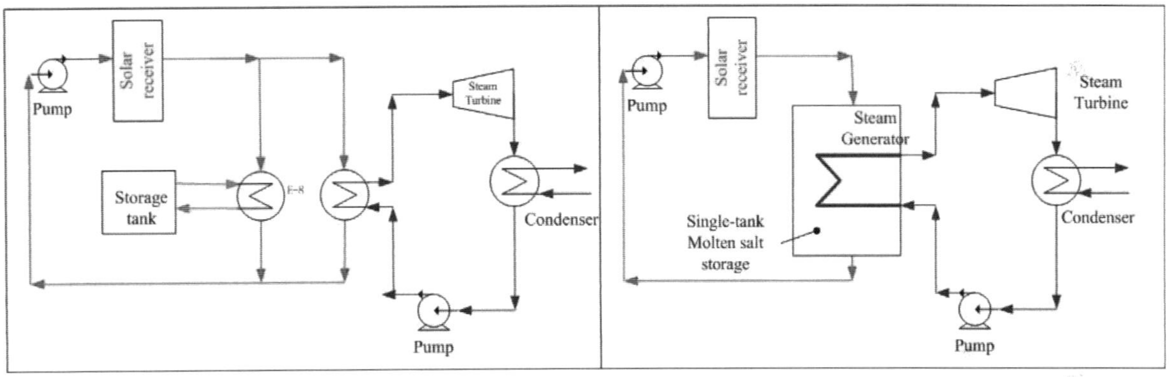


Figure 1 Thermal storage base case and the improved design process

**(703f) Thin-Film Composite Membranes for Osmotic Power Generation**

Li, X., National University of Singapore
Chung, T. S., National University of Singapore
Ong, R. C., National University of Singapore
Originally presented on: 11/7/2013 14:35:00 - 15:00:00

**Thin-Film Composite Membranes for Osmotic Power Generation**

Xue LI[a], Tai-Shung CHUNG-[a, b,*]

[a]NUS Graduate School for Integrative Science and Engineering, National University of Singapore, Singapore 117456, Singapore

[b]Department of Chemical & Biomolecular Engineering, National University of Singapore, Singapore 117576, Singapore *Corresponding Author's E-mail: chencts@nus.edu.sg **Keywords:** Osmotic power generation; Thin-film composite membranes; Pressure retarded osmosis. **Abstract**

Pressure retarded osmosis (PRO) is a membrane-based technology that harvests the energy of mixing two solutions from different salinity and converts it to electricity with the aid of a semi-permeable PRO membrane and a hydro-turbine. In addition to utilize seawater and river water as the feed pair, the PRO process can use highly concentrated RO retentate and recycled municipal wastewater as the feed pair for power generation. By doing so, not only can the environmental issue such as the disposal of RO retentate be solved, but also produce much higher osmotic energy. Therefore, the development of PRO membranes for osmotic power generation will have repercussion effects on the production of both clean energy and clean water.

As the heart of PRO processes, more studies on PRO membranes are expected to further accelerate the process in the near future. The thin-film composite (TFC) membranes, a commonly used PRO membrane nowadays, consist of a TFC layer of cross-linked polymers which acts as the primary barrier to salt and is responsible for giving a reasonable selectivity. In PRO processes, the draw solutions are pressurized by high hydraulic pressures to later push the hydroturbine and convert osmotic power into electrical power. However, under high pressures, membranes inevitably lose selectivity because of deformation and damage of both TFC layer and supporting layer. In our work, a systematical investigation has been demostrated on the evolution of morphological deformation and membrane performance in terms of compacted structure, water flux, reverse salt flux, and power density as a function of hydraulic pressure for membranes used in high-pressure PRO processes. In addition, an insight to PRO membranes will be given on the effects of membrane modifications and post-treatments.

**Acknowledgments**

This research was funded under the project entitled ?Membrane development for osmotic power generation, Part 1. Materials development and membrane fabrication? (1102-IRIS-11-01) and NUS grant number of R-279-000-381-279. This research grant is supported by the Singapore National Research Foundation under its Environmental & Water Technologies Strategic Research Programme and administered by the Environment & Water Industry Programme Office (EWI) of the PUB. Special thanks are due to Dr. Panu Sukitpaneenit, Dr. Sui Zhang, and Miss Xiuzhu Fu for their valuable suggestions and kind help.

**(707b) CO2 Capture and Conversion Based On ILs**
Zhang, S., Institute of Process Engineering, Chinese Academy of Sciences
Originally presented on: 11/7/2013 12:55:00 - 13:20:00

## $CO_2$ Capture and Conversion Based on ILs

Suojiang Zhang*, Jian Sun

*Institute of Process Engineering, Chinese Academy of Sciences,100190, Beijing, PR China*

**Corresponding author: sjzhang@home.ipe.ac.cn*

An ionic liquid based $CO_2$ capture technology has been developed. To enhance the absorption efficiency, a series of ILs were synthesized and screened including amino functionalized IL.[1] The performance of IL mixture for $CO_2$ capture was also carried out.[2]The system integration result indicated that the new process resulted in about 27% reducing of energy consumption comparing to the traditional amine process.

As a way out for $CO_2$ capture, the synthesis of cyclic carbonates via cycloaddition of $CO_2$ to epoxides is studied, where the synthesized ILs are found to be very efficient catalysts for the cycloaddition reaction. Our previous work indicated that the hydroxyl group in solvent, IL or carrier all could play a promotion role on the reaction. [3] Based on the fundamental understanding, hydroxyl or carboxyl-functionalized ILs were developed and found to be much more active.[4] Furthermore, these ILs are chemically immobilized on solid supports, including silica, resin, zeolites, and chitosan, which show high stability and efficiency.[5]

### Refferences

[1] a) X. P. Zhang, X. C. Zhang, H. F. Dong, Z. J. Zhao, S. J. Zhang, Y. Huang. ***Energy Environ. Sci***., 2012, 5: 6668-6681. b) J. M. Zhang, J. Sun, X. C. Zhang, Y. S. Zhao, S. J. Zhang. ***Greenhouse Gas Sci Technol***., 2011, 2:142-159.

[2] a) Y. S. Zhao, X. P. Zhang, Y. P. Zhen, H. F. Dong, G. Y. Zhao, S. J. Zeng, X.Tian, S. J. Zhang. ***Int. J. Greenh. Gas Con***., 2011, 5:367-373; b) Y. S. Zhao, X. P. Zhang, H. F. Dong, Y. P. Zhen, G. H. Li, S. J. Zeng, S. J. Zhang. ***Fluid Phase Equilib***., 2011,302: 60-64.

[3] a) J.Q. Wang, J. Sun, W.G Cheng, S.J. Zhang. ***Phys. Chem. Chem. Phys.***, 2012, 14, 11021-11026; b) J. Sun, J. Y. Ren, S. J. Zhang, W. G. Cheng. ***Tetrahedron Lett***., 2009, 50(4): 423-426.

[4] a) X. Chen, J. Sun, J. Q. Wang, W. G. Cheng. ***Tetrahedron Lett***., 2012, 53: 2684-2688; b) J. Sun, L. J. Han, W. G. Cheng , J. Q. Wang, X. P. Zhang, S. J. Zhang. ***ChemSusChem.***, 2011, 4(4): 502-507; c) J. Sun, S. J. Zhang, W. G. Cheng, J. Y. Ren. ***Tetrahedron Lett.***, 2008, 49(22): 3588-3591.

[5] a) W. G. Cheng, X. Cheng, J. Sun, J.Q. Wang, S. J. Zhang. ***Catal. Today***, 2013, 200:117-124; b) J. Sun, J.Q. Wang, W. G. Cheng, J. X. Zhang, X. H. Li, S. J. Zhang, Y. B. She. ***Green Chem***., 2012, 14: 654-660. c) J. Sun, W. G. Cheng, W. Fan, Y. H. Wang, Z. Y. Meng, S. J. Zhang. ***Catal. Today***, 2009, 148: 361-367.

### Acknowledgments

This work was supported by National Basic Research Program of China (2009CB219901), and National Science Fund of China (21006117).

**(707c) Modeling of CO2 Capture Using Microalgae**

Andrade, L. M., University of Sao Paulo
Mendes, M. A., University of Sao Paulo
Silveira, C. A.,
Oller Nascimento, C. A., University of SÃ£o Paulo
Originally presented on: 11/7/2013 13:20:00 - 13:40:00

In order to model the growth of microalgae in a open-pond reactor, a 1000L system has been built and operated under realistic outdoor conditions. Environmental data (light intensity, temperature, pH) were captured with a datalogger and the growth was measured. A phenomenological mathematical model was built for the open reactor. The comparison of the simulated data to the experimental data shows good agreement.

**(707d) Biogas Sweetening Via a Carbamate Crystallization Approach**

Zheng, S., ZhejiangUniversity
Liu, Q.,
He, Y.,
Shi, Y.,
Originally presented on: 11/7/2013 13:40:00 - 14:00:00

The $CO_2$ in biogas significantly limited its potential applications. The regeneration process of aqueous alkanolamine based $CO_2$ capture often requires a large amount of energy, which makes the process costly. In order to enhance the energy efficiency of the capture, a carbamate crystallization approach was proposed in this study where water was replaced with non-aqueous solvents. It was observed that triethylenetetramine (TETA) carbamate crystallize out as supernatant solid from ethanol solution upon the adsorption of $CO_2$. A TETA/non-aqueous solvent solution shows higher absorption capacity and absorption rate than alkanolamine-water solution. The mole ratio of $CO_2$/TETA in carbamate was determined to be around 2:1 by elemental analyzer. The regeneration capabilities of solid carbamates was also investigated. Thermal behavior of the carbamate was examined with differential scanning calorimetry and thermogravimetric analysis. It was shown that the decomposition of TETA-carbamate commenced at about 90°C. The separation of carbamate as solid phase can offer great advantage by letting less volume to regenerate and make this approach a promising alternative for biogas sweetening.

**(707e) Preparation and Absorption Behaviors Study of Doped Li4SiO4 At Low CO2 Concentration**
Qi, Z., East China University of Science and Technology
Sai, Z., East China University of Science and Technology
Originally presented on: 11/7/2013 14:00:00 - 14:20:00

## Preparation and Absorption Behaviors Study of Doped $Li_4SiO_4$ at Low $CO_2$ Concentration

Sai Zhang, Qi Zhang*,Qian Ye, Haiyan Wang, Zibin Zhu

Dept. of Chemical Engineering, East China University of Science and Technology, Shanghai 200237, China

**Abstract**

Different K-doped $Li_4SiO_4$ samples were prepared to improve the pure $Li_4SiO_4$ adsorbent's sorption/desorption properties in the critical SESMR practical conditions with low $CO_2$ partial pressure and steam atmosphere. The optimized sample was analyzed by sorption experiments and simulation with FactSage5.5 Software. The results show that the K-doped $Li_4SiO_4$ has a better $CO_2$ absorption ability than pure $Li_4SiO_4$ as the K can form a eutectic compound with $Li_4SiO_4$ which enhances the $Li^+$ and $O^{2-}$ ions diffusion and thus facilitating the $CO_2$ absorption ability of $Li_4SiO_4$. It is also found that steam not only enhances the $Li^+$ mobility but also reacts with $Li_4SiO_4$, therefore increase the mass uptake of the adsorbent. In order to make the adsorbent more practical, sorption-desorption cycle processes were also carried out. The results show that the optimized $Li_4SiO_4$ has a lower desorption temperature and stable absorption ability.

Furthermore, the sorbents, $Li_4SiO_4$ and different K-doped $Li_4SiO_4$, were evaluated using a bench scale unit, of which the results can provide reference for the subsequent industrial design. $Li_4SiO_4$ and different K-doped $Li_4SiO_4$ were prepared by solid-state reaction method. The results was obtained from the studies of different $Li_4SiO_4$ samples in a fixed-bed reactor with 15%$CO_2$ and 85%$N_2$ at 575Â°Ã¦. The pure $Li_4SiO_4$ showed a short breakthrough time, approximately 20 min, while modified $Li_4SiO_4$ samples presented a relative 20~40 min longer breakthrough time under same conditions. The results of the experiment indicate that doping K into $Li_4SiO_4$ can improve the adsorption capacity of the sorbents. Comparing with MgO-based $K_2CO_3$[1] and hydrotalcites[2] sorbents, which breakthrough time are no more than 40min and 13min, respectively, the relatively long breakthrough time of K-doped $Li_4SiO_4$ exhibits huge potential for industrial applications. K-$Li_4SiO_4$ with good performance at low $CO_2$ partial pressure can be applied into sorption enhanced steam methane reforming system, in which the $CO_2$ concentration can be also as low as 15%. Meanwhile, the regeneration period can be prolonged by doping appropriate amount K into Li4SiO4, which means cost for recycling can be reduced as a result. K-doped Li4SiO4 has a certain value in industrial application, both practicability and economically.

**Key words**: $CO_2$ acceptors, Lithium orthosilicate, on-line absorption behaviors

**References**

[1]Soo Chool Lee, Ho Jin Chae, et al. Development of Regenerable MgO-Based Sorbent Promoted with K2CO3 for CO2 Capture at Low Temperatures[J]. Environ.Sci. Technol, 2008, 42: 2736-2741

[2]Hendricus Th. J. Reijers, Sacha E. A.Valster-schiermeier, et al. Hydrotalcite as CO2 Sorbent for Sorption-Enhanced Steam Reforming of Methane[J]. Ind. Eng. Chem. Res, 2006, 45, 2522-2530

**(707f) Zr Doping Effect On CeO2 As the Catalyst in the CO2 â€“ Ethylbenzene Reaction**

Wang, Q.,
Li, W., Taiyuan University of Technology
Originally presented on: 11/7/2013 14:20:00 - 14:40:00

Under the thermodynamic limitation , dehydrogenation of ethylbenzene (EB) to styrene (ST) with $CO_2$ is widely investigated. The lattice oxygen mobile capacity of catalyst is the important influencing factors in the $CO_2$-EB reaction. $CeO_2$ of the $CaF_2$ structure with the excellent oxygen storage/release capacity (OSC) is preferred as the catalyst. The addition of $Zr^{4+}$ to the cubic structure of $CeO_2$ can greatly increase the OSC and improve its surface area, thermal stability, mechanical strength and surface acidity. The Zr content of $Ce_xZr_{1-x}O_2$ solid solutions strongly influence the crystal structure, the mobility of lattice oxygen and the catalytic performance. $Ce_xZr_{1-x}O_2$ solid solutions replacing the pure $CeO_2$ as the catalyst could be used in oxidative dehydrogenation of EB to ST with $CO_2$ for changing the lattice structure of $CeO_2$ and improving the transmission capacity of lattice oxygen.

In this work, $Ce_xZr_{1-x}O_2$ solid solutions with different Zr proportions were synthesized using hydrothermal method. The synthesized catalysts were characterized by X-ray power diffraction pattern, Raman spectroscopy and $N_2$-adsorption. It was found that the catalysts were solid solution mixture having specific surface area more than 100 $m^2/g$. By the means of the $H_2$ temperature-programmed reduction, the lattice oxygen mobile capacity was calculated, corresponding to the order is $Ce_{0.5}Zr_{0.5}O_2 > Ce_{0.7}Zr_{0.3}O_2 > Ce_{0.3}Zr_{0.7}O_2 > CeO_2$. The performance of the catalysts for dehydrogenation of EB to ST was evaluated in a fixed-bed reactor under the conditions 823 K, 1 atm, and 20 of the $CO_2$/EB molar ratio. Results showed that the lattice oxygen mobile capacity of $Ce_xZr_{1-x}O_2$ influenced the catalytic activity and promoted activity of $CO_2$. The $Ce_{0.5}Zr_{0.5}O_2$ showed the highest activity with the EB conversion of 50% and the ST selectivity above 96%. The mechanistic model of EB dehydrogenation with $CO_2$ over the $Ce_xZr_{1-x}O_2$ catalysts was proposed.

**(707g) Combined Natural Gas Liquefaction and Power Generation Plant With Carbon Dioxide Sequestration**

Verma, S., Schlumberger Doll Research
Ramakrishnan, T. S.,
Originally presented on: 11/7/2013 14:40:00 - 15:00:00

It is generally well accepted that an increase in anthropogenic emissions is a leading cause of the increase in levels of carbon dioxide ($CO_2$), one of the greenhouse gases, in the atmosphere. Generation of electrical power is responsible for a roughly 40% of these emissions. Design of new zero emission power plants is critical in any effort to reduce emissions of $CO_2$.

We discuss the design of a combined natural gas liquefaction and power generation plant with no atmospheric emissions of greenhouse gases. More specifically, this design relates to (i) a zero emission process for power generation as the $CO_2$produced during the power generation cycle is sequestered into a terrestrial formation and (ii) a process by which the high pressure natural gas stream is expanded to provide refrigeration and power. The recovered energy may be used to liquefy part of the natural gas stream to provide pure liquefied natural gas (LNG).

This plant uses natural gas from a well as a fuel. The high-pressure stream is liquefied in a reverse-Brayton cycle and the remainder of the natural gas stream is fed to a solid oxide fuel cell to produce power. The effluent of the solid oxide fuel cell is completely combusted to reduce it to $CO_2$ and water. This stream is subsequently cooled to remove the condensed water.

Energy recovered in expansion of natural gas is used for CO2 stream compression, reducing the energy required to by 13.7%. Alternatively, if LNG is not desired as a product, then as much as 30% of the compression energy may be supplied from expansion of natural gas. Overall power plant efficiencies of 50 – 60% are possible. About 27.5% wt/wt of the incoming natural gas stream may be liquefied in this cycle.

**(721a) Evaluation of Microbial Lipopeptide Biosurfactants for Oil-Spill Remediation**

Marti, M. E., Iowa State University
Colonna, W., Iowa State University
Partha, P.,
Zhang, H.,
Green, C.,
Reznik, G., Modular Genetics, Inc.
Pynn, M., Modular Genetics, Inc.
Jarrell, K., Modular Genetics, Inc.
Nyman, A.,
Somasundaran, P., Columbia University
Lamsal, B., Iowa State University
Glatz, C. E., Iowa State University
Originally presented on: 11/7/2013 15:15:00 - 15:40:00

The selection of a surfactant is influenced by its efficacy and environmental impact and may vary depending on the application. There is an increasing demand for biodegradable and non-toxic surfactants in oil spill remediation. Currently, surfactants are mainly derived via chemical synthesis from petroleum-based hydrocarbons. Lipopeptide biosurfactants, produced by microorganisms, are favored for their attractive surface-activity, biocidal and physiological properties. In the present study, the properties of two lipopeptide biosurfactants and several commercially-available synthetic surfactants were compared. Both biosurfactants, surfactin and FA-Glu, were produced using genetically-modified strains of Bacillus subtilis on glucose-mineral salts media. With both surfactants, rates of synthesis slowed after ~24 h with depletion of the carbon source. Typical titers from shake flasks were ~200-400 mg/L and ~5.5-6.0 g/L, for FA-Glu and surfactin respectively. When they were produced in large fermentors, the surfactants were recovered via foam fractionation and were enriched several fold over surfactant levels in the residual broth. The initial growth rate of FA-Glu producing cells was faster than that of surfactin-producing cells. The surfactants were recovered at purities of >98% and used in physico-chemical and toxicity experiments. Critical micelle concentrations (CMC) of surfactin and FA-Glu were 0.015 and 0.10 g/L, respectively, in distilled water, and were comparable to that of Corexit 9500, a commercially-available synthetic surfactant. Higher CMCs and lower dispersant-to-oil ratio (DOR) values were observed in saline. This was most probably due to the effects of electrolyte and pH on these properties. Both biosurfactants were more effective oil dispersants than Corexit in fresh water. However, in saline, surfactin was most effective, while Corexit was superior to FA-Glu. Fish survival trials showed that surfactin was the most toxic, while FA-Glu had the lowest toxicity of the surfactants tested. Toxicity increased with elevated saline levels; however, with FA-Glu, the effect of saline on toxicity was minimal.

## (721b) Feasibility Study of Production of Methanol and Dimethyl Ether From Flare Gas

Fukuda, S., Japan Petroleum Exploration CO.,LTD.
Chaki, K., Japan Petroleum Exploration CO.,LTD.
Wakatsuki, T., Japan Petroleum Exploration CO.,LTD.
Originally presented on: 11/7/2013 15:40:00 - 16:05:00

We studied the feasibility of effectively utilizing gas associated with oil production (APG), including the economics of producing methanol and dimethyl ether from APG. Detailed estimates indicate that the internal rate of return (IRR) of these processes is >10%, indicating potential for success.

When crude oil is produced, APG comes to the surface from the oil reservoirs. In most cases, this APG is used as generator fuel onsite or transported by pipeline and sold elsewhere. However, in some areas far from gas markets, much of the APG is ignited and released into the atmosphere due to the high cost of constructing pipelines. In remote areas and/or at small-scale oil fields, it is difficult to recover construction costs through sales. Globally, APG comprises 150 billion $m^3$/yr, equivalent to 25% of the USA's gas consumption. $CO_2$ emissions due to flaring of APG are 400 billion tons/yr, equivalent to about 1.2% of global $CO_2$ emissions. $CO_2$emissions have serious consequences to the environment; thus, reducing the amount of APG flared at oil fields without using pipelines is a priority.

Heavy hydrocarbons such as liquid petroleum gas (LPG) can be separated by cooling, compression, or absorption. Separated gases can be liquefied easily and transported by truck. Considering the cost of constructing a pipeline, we assumed that transporting the gas by truck was more feasible. If all of the APG could be separated and trucked, construction of a pipeline would not be needed. However, APG consists of light hydrocarbons, which are not easy to liquefy through physical phase changes. Thus, we considered chemical reactions to liquefy APG.

Methanol and dimethyl ether are liquid chemicals produced from hydrocarbons. Methanol is used in the manufacture of various chemicals and liquid fuels and is globally in demand. On the other hand, the market for dimethyl ether is not currently large. However, it has the potential to increase, because dimethyl ether is expected to become a substitute for LPG. Accordingly, methanol and/or dimethyl ether could be manufactured from APG and trucked to market. We developed a process flow based on the concepts above and the actual conditions at one oil field. Production of methanol requires producing syngas from hydrocarbons through various possible reactions. We evaluated the production of syngas through steam reforming and partial oxidation. Thus, we considered the following four cases consisting of two reactions for producing syngas and two final products:

Case A: Methanol synthesis via syngas production by partial oxidation

Case B: Dimethyl ether synthesis via syngas production by partial oxidation

Case C: Methanol synthesis via syngas production by steam reforming

Case D: Dimethyl ether synthesis via syngas production by steam reforming

We estimated the feasibility of these processes based on operating expenses, capital expenses, and the actual conditions at one oil field. The price of the ingredient gas was considered to be $0 US, because the chemicals are produced from the oil field's own APG. The results of the economic analysis indicated that the IRR of these processes was >10% and thus they had the potential for success. Comparison of the four cases indicated that use of partial oxidation was superior to use of steam reforming for larger facilities, because the reactor for partial oxidation is smaller than that for steam reforming. Considering the IRR for dimethyl ether production, Case B or Case D may be successful depending on the size of the market. The comparison reveals the feasible conditions for each case. The effects of conditions and details of the process flow are discussed.

## (721c) Abatement of Greenhouse Gases Into Methanol in a Ntp-DBD Reactor - Effects of Experimental Parameters

Mahammadunnisa, S., Indian Institute of Technology-Hyderabad
Manoj Kumar Reddy, P.,
Subrahmanyam, C., Indian Institute of Technology-Hyderabad
Originally presented on: 11/7/2013 16:05:00 - 16:30:00

**Abatement of Greenhouse gases into Methanol in a NTP-DBD reactor**

**- Effects of Experimental Parameters**

**Abstract:** Direct conversion of natural gas to liquid products such as methanol and formaldehyde has been receiving a great attention [1, 2]. However, converting methane ($CH_4$) to methanol ($CH_3OH$) with desired selectivity is difficult as in general, the oxidation proceeds to complete to carbon oxides (CO or $CO_2$) and water in conventional methods [2]. On the other hand, utilization of nitrous oxide ($N_2O$), as a co-reactant for $CH_4$ conversion appears to be very promising. Methane conversion by $N_2O$ may also beneficial due to potential green-house nature of these two reactants (The GWP for $CH_4$, $N_2O$ is 21 and 280 times more heat-trapping than that of $CO_2$). In recent years, industrial applications of non-thermal plasma (NTP) generated under ambient conditions has been receiving a great interest [3-5]. NTP generated by electrical discharges under ambient conditions has specific advantages like mild operating conditions, as NTP is composed of energetic electrons that may initiate the reaction, ions, excited atoms, radicals and meta stable ions. These energetic electrons initiate the chemical reactions in gas phase. With this back ground, the present study is aimed at co-processing of two green house gases $CH_4$ and $N_2O$ for the production of oxygenated products like methanol and formaldehyde.

InÃ¯Â¬?uence of feed gases proportion on oxidation of methane ($CH_4$) with nitrous oxide ($N_2O$) has been studied in dielectric barrier discharge (DBD) plasma operated under ambient conditions (Fig.1). Plasma reactor was powered by a high voltage AC generator (Yaskawa varispeed F7 AC inverter upto 40 kV and 50 Hz). The gas mixture was controlled by a set of mass Ã¯Â¬?ow controllers (Aalborg-USA). $CH_4$ (10 vol% diluted in Ar) and $N_2O$ (5 vol% diluted in Ar) gases were introduced into the reactor. At the reactor outlet gas analysis was made with an online gas chromatograph (Varian 450-GC), whereas an on-line GC-MS (Thermo Fischer) was used to identify the by-products. Typical results indicated the formation of methanol ($CH_3OH$), formaldehyde (HCHO), hydrogen ($H_2$), carbon monoxide (CO) and carbon dioxide ($CO_2$).

The important findings of the present study may be summarized as

1. NTP is effective in direct decomposition of $N_2O$ into $N_2$ and O that will be alter converted to $O_2$ in the absence of a catalyst.
2. Increasing $CH_4/N_2O$ molar ratio in the feed gases favors the reagents conversion and consequently promotes the formation of each by product.
3. Co-processing $CH_4$ and $N_2O$ produces $CH_3OH$ and HCHO, however, the selectivity to these products is rather low.
4. Water vapor has a positive effect on selectivity.

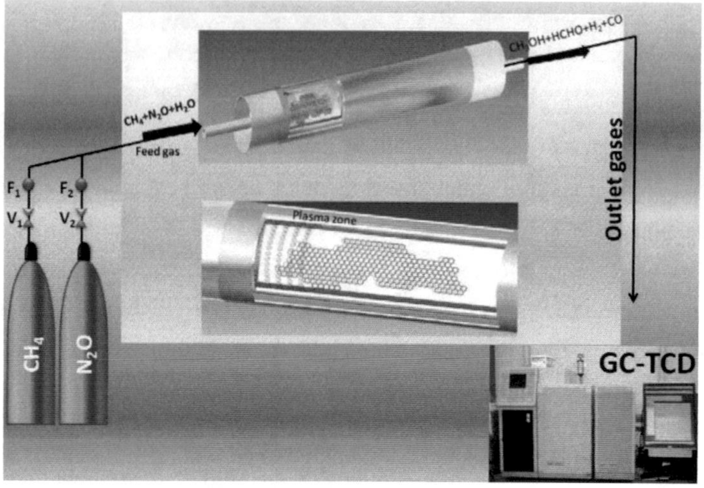


Figure 1. Schematic representation of NTP-DBD reactor for oxidation of methane ($CH_4$) with nitrous oxide ($N_2O$)

**References:**

1. [1] L.B. Han, S. Tsubota, T. Kobayashi, M. Haruta, Formation of methanol by the gas-phase partial oxidation of methane under normal pressures, J. Chem. Soc. Chem. Commun. (1995) 93--94.

[2] L.M. Zhou, B. Xue, U. Kogelschatz, B. Eliasson, Partial oxidation of methane to methanol with oxygen or air in a non-equilibrium discharge plasma, Plasma Chem. Plasma Process. 18 (3) (1998) 375--393.

[3] K. Okazaki, S. Hirai, T. Nozaki, K. Ogawa, K. Hijikata, Plasma chemical reactions at atmospheric pressure for high efficiency use of hydrocarbon fuels, Energy 22 (2--3) (1997) 369--374.

[4] Sk. Mahammadunnisa, E.L. Reddy, P. M. K. Reddy and Ch. Subrahmanyam., A facile approach for direct decomposition of nitrous oxide assisted by nonthermal plasma, Plasma Processes and Polymers. (DOI: 10.1002/ppap.201200114)

[5] Sk. Mahammadunnisa, P. M. K. Reddy, and Ch. Subrahmanyam., dry reforming assisted non-thermal plasma catalytic reactor, ACS Energy & Fuels (dx.doi.org/10.1021/ef302193e).

**(721d) Progress and Economic Considerations for the Biological Production of Triterpene Biofuels From Gases**

Khan, N. E., Penn State University
Myers, J. A., Penn State University
Johnson, R., Penn State University
Rajangam, A., The Pennsylvania State University
Nybo, E., University of Kentucky
Chappell, J., University of Kentucky
Curtis, W., The Pennsylvania State University
Originally presented on: 11/7/2013 16:30:00 - 16:55:00

We present our progress in producing triterpene fuel precursors from $CO_2$, $H_2$ and $O_2$as part of the ARPA-E electrofuels program. While increasing product titer provides an obvious improvement to process economic feasibility, we also present that the maintenance energy of bacteria, or the energy needed to keep it alive, is a significant economic consideration for fuel production.

The Advanced Research Projects Agency (ARPA) program seeks to produce liquid transportation fuels from renewable electricity and $CO_2$. Botryococcenes ($C_{30}$ hydrocarbons) from *Botryococcus braunii* is a very good potential fuel molecule for this purpose due to its high energy density, hydrophobicity and similarity to petroleum crude. A primary research milestone is the genetic engineering of the isoprene metabolic pathway through isopentenyl diphosphate (IPP) to triterpenes. This is being accomplished using the genes from Mevalonate and MEP pathways as well as triterpene synthase genes from an algae, *Botryococcus braunii,* known to produce hydrocarbons. We were able to use the vector systems that have been used earlier to transform other *Rhodobacter*species with constitutive promoters (pLac) and native promoters to successfully produce isoprenes. Current efforts are undertaking a "breadth screen" for different vectors, promoters, knock-in and hosts to examine the "landscape" for triterpene production. A "depth" screen is underway to better understand the rate limiting enzymes and metabolic fluxes. The current status of these screens will be presented.

We are using economic considerations of the envisioned process to help drive research priorities in addition to the goal of genetically engineering the botryococcenes pathway from *B. braunii* into a chemolithoautotrophic bacteria – *Rhodobacter capsulatus.* Since the process involves the production of a moderate value fuel molecule from useful gaseous substrates ($H_2$ and $CO_2$), the growth of the bacteria has to be minimized to maximize the yield to to final product. Therefore the maintenance coefficient of the bacterial host becomes important as this is the absolute minimum energy wasted in the process without generating any product. To this end, we have undertaken the measurement of the growth yield and maintenance coefficient of *Rhodobacter capsulatus* and *Ralstonia eutropha* (another well-studied autotrophic production host) and determine their relative impact on the overall economics of the process. While it is relatively straight-forward to close a mass balance in the liquid phase, it is quite difficult to do it accurately in the gas phase because of effects of temperature, pressure and composition on the measurements. As such, literature references for growth yield and maintenance coefficient of autrophic bacteria are scarce. We have carried out chemostat studies under various conditions using an elaborate setup to continuously measure the inlet and outlet gas flow rate and composition from the reactor to ensure accurate determination of the parameter values. Our results indicate that under oxygen limited conditions the maintenance coefficient of *R. capsulatus* is significantly lower than that of *R. eutropha*. On the other hand, the growth yield of *R. eutropha* is much better than *R. capsulatus*. This represents critical trade-offs in terms of process design. The other important parameter affecting the process is the gas-liquid mass transfer of the bioreactor employed as this would affect the volumetric productivity of the process and thus the capital cost. We have carried out a detailed process design and economic analysis (incorporating operating, fixed and capital costs at large industrial scale) to determine the overall feasibility of the process. There exist multiple routes of going from gaseous substrates to liquid fuels. We will present the results of the different economic scenarios incorporating both literature and experimentally obtained values for the organism and reactor performance. This analysis will be used to draw conclusions about the most probable path forward for this technology.

The results of this project thus far challenge the typical paradigm of working with organisms that are inherently fast growing and easy to culture. We suggest that as genetic engineering capabilities extend to more difficult organisms, some of the more fringe 'tougher, more efficient' organisms may be better suited for fuels production than their fast-growing 'easy to manipulate' counterparts.

## (727a) Effect of Pore and Particle Size of Oxygen Carriers On Chemical-Looping Combustion and Reforming

Han, L., University of Connecticut
Zhou, Z., University of Connecticut
Bollas, G. M., University of Connecticut
Originally presented on: 11/7/2013 15:15:00 - 15:36:00

### Effect of pore and particle size of oxygen carriers on chemical-looping combustion and reforming

**Lu Han**, Zhiquan Zhou, George M. Bollas

### Abstract

Chemical-looping combustion (CLC) addresses the current $CO_2$ problem caused by fossil fuel utilization for power production. The main advantage of CLC over conventional power production technologies is that direct contact between air and fuel is circumvented, so that a pure stream of $CO_2$ is obtained without nitrogen dilution. An intermediate oxygen carrier (metal/metal oxide) is alternatively oxidized and reduced by the fuel and oxygen. Two successive reactions take place between two reactors shown in **Figure 1**. In this manner, $CO_2$ can be captured from the flue gases of the Fuel reactor without additional separation and high energy penalty. Chemical-looping reforming (CLR) is similar to CLC however a mixture of steam and fuel is sent to the Fuel Reactor to induce reforming reactions and produce syngas.

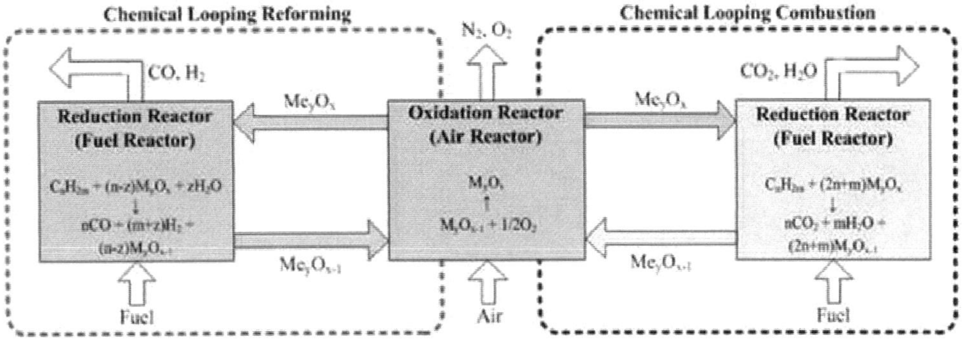


Figure 1: Basic concept of chemical-looping combustion and reforming processes

Research in the chemical-looping (CL) process has been focused on the development and testing of oxygen carrier particles. Favorable characteristics of high performance oxygen carriers include: high reactivity toward oxidation and reduction cycles, high melting temperature, low tendency of attrition and fragmentation, low toxicity, low cost and high combustion efficiency. Most reported work on CL has been accomplished using oxygen carrier particles where the active component (Cu, Fe, Ni, Mn) is formulated on an inert material [1]. This work focuses on reported studies utilizing Ni-based oxygen carriers for its high reactivity in the combustion of methane and catalytic properties.

A particle model is developed to study the effect of internal transport on the reaction rates. Intraparticle diffusion limitations need to be accounted for particles that are millimeters in size [2]. Due to the importance of Knudsen diffusion, internal transport resistance of different gases gives rise of concentration gradients inside the pores. Reaction rates are observed highest at the particle surface and decrease along the pore length. This knowledge can aid in oxygen carrier design to avoid over-estimation of the system performance when no mass transfer effects are considered. Incorporated into a reactor model, optimum particle dimensions are proposed to conduct CLR in a fixed-bed reactor that makes use of the intraparticle effects to maximize $H_2$ selectivity.

It is also worth to explore the change in particle morphology over many reduction and oxygen cycles in CLC and CLR. Investigators have reported increasing reactivity with usage associated with micro-cracking of the particle surface and migration of the Ni particles to the surface. With the particle model, the real changes to particle porosity, surface area, and pore size are accounted for, which influence the role of transport resistance on the performance. Observations made from experiments on the particle morphology are utilized to predict reactivity changes over cyclic tests.

Lastly the interaction between NiO and $Al_2O_3$ support is addressed because when present, the spinel can significantly lower the reactivity of the oxygen carrier. For instance, the reactivity of two supported NiO-oxygen carriers differing only by the calcination temperature of the support was measured in a thermogravimetric analyzer at 800Â°C in a $H_2$-environment (**Figure 2**). The $Al_2O_3$ support existed in the γ-phase in the low temperature pre-treatment and exhibited a phase transformation into α upon high temperature calcination. More instance of spinel formation occurred in the γ-$Al_2O_3$ support which resulted in comparably lower reactivity of the oxygen carrier. A kinetic scheme that combines the simultaneous reduction of NiO and $NiAl_2O_4$ is used in the particle model. Reduction data of Ni-based oxygen carriers with varying spinel phases are studied for the purpose of assessing the impact of spinel on kinetic rates. Kinetic parameters derived from the particle model are presented and model-experiment deviations are discussed.

**Acknowledgement**: This material is based upon work supported by the National Science Foundation under Grant No. 1054718.

**References:**

[1] M. Johansson, T. Mattisson, M. Ryden, and A. Lyngfelt, ?Carbon Capture via Chemical-Looping Combustion and Reforming Chemical-Looping Combustion,? in *International Seminar on Carbon Sequestration and Climate change*, 2006, no. October, pp. 24?27.

[2] S. Noorman, F. Gallucci, M. van Sint Annaland, and J. a. M. Kuipers, ?A theoretical investigation of CLC in packed beds. Part 1: Particle model,? *Chemical Engineering Journal*, vol. 167, no. 1, pp. 297?307, Feb. 2011.

**(727b) Thermal Evaluation of Chemical Looping Process for Hydrogen Production Using Bivalent Metal Added Iron Oxide As An Oxygen Carrier**

Cho, W., Korea Institute of Energy Research
Kang, K., Korea Institute of Energy Research
Park, C., Korea Institute of Energy Research
Jeong, S. U., Korea Institute of Energy Research
Lee, D., KAIST
Seo, M. W., Korea Institute of Energy Research
Kim, C. H., KIER
Bae, K. K., Korea Institute of Energy Research
Kim, S. D., KAIST
Originally presented on: 11/7/2013 15:36:00 - 15:58:00

The chemical looping process for hydrogen production co-generates hydrogen and electricity with simultaneous capture of carbon dioxide. This process is composed of interconnected three reactors: a fuel reactor where an oxygen carrier is reduced by fuel, a steam reactor where hydrogen is produced by water splitting, and air reactor where air is combusted. The iron oxide is used as a suitable oxygen carrier from thermodynamic and chemical equilibrium point of view. The many studies were conducted to improve the reactivity of the iron oxide such as preparation methods, sintering temperatures, supporters, etc. In addition, the mixing the bivalent metal (M = Cu, Ni, Mo) to the iron oxide has been studied to enhance the reactivity of the redox reaction. In this study, the bivalent-added iron oxides showed the kinetic improvement, higher steam conversion to hydrogen, and the lower carbon deposition. The chemical looping process was simulated and the thermal evaluation was conducted based on the experimental data, varying the oxygen carriers. The several cases were compared varying the temperatures and the extent of solid conversion in the fuel and steam reactors.

**(727c) Modeling of Cu Oxidation in Air Reactors of Chemical Looping Combustion Systems**

Goldstein, E. A., Stanford University
Mitchell, R. E., Stanford University
GÃ¼r, T. M., Stanford University
Originally presented on: 11/7/2013 15:58:00 - 16:19:00

An oxidation model is developed that describes *Cu* (copper) oxidation kinetics to *CuO* (cupric oxide) for applications in the air reactor of chemical looping combustion (*CLC*) systems. The model, which is based on the Wagner theory of oxidation, is developed by considering the fluxes of interstitial and vacancy defects in the cation sublattice of $Cu_2O$ (cuprous oxide) and *CuO*. Defects in the anion sublattice are a priori included in the model, however are subsequently found to be too small in concentration and diffuse too slowly to contribute significantly to the oxidation of Cu. The oxidation model presented here is capable of predicting Cu oxidation rates and relative thicknesses of the resulting $Cu_2O$ and *CuO* oxide layers as a function of the partial pressure of oxygen in the ambient and temperature.

Experiments were performed in a thermogravimetric analyzer (*TGA*) at conditions relevant to air reactors of *CLC* systems to obtain data that was used to characterize parameters in the model. Experiments were performed on copper spheres having diameters in the range 15 to 25 μm, at temperatures between 500 and 900 °C and at 1 atm. The gaseous environments contained 0.3%, 6% and 21% $O_2$, by volume, with the balance $N_2$. Based on this model, the temperature and oxygen partial pressure dependencies of the oxidation rate as well as overall rate constants were determined and compared to published results for Cu oxidation. A sensitivity analysis was performed in order to assess the influence of defect diffusivities and formation energies as well as particle diameter and surface area on oxidation rates. Based on our understanding of defects and their role in Cu oxidation, insights regarding the design of copper particles for use in CLC have been made. Assuming that oxidation is the rate-limiting process in *CLC*, the recirculation rate and copper loadings have also been estimated when CuO is used to oxidize a typical coal.

**(727d) An Overview of Chemical-Looping Reduction in Fixed-Bed and Fluidized Bed Reactors Focused On Oxygen Carrier Utilization and Reactor Efficiency**

Zhou, Z., University of Connecticut
Han, L., University of Connecticut
Du, S., University of Connecticut
Fischer, A., University of Connecticut
Bollas, G. M., University of Connecticut
Originally presented on: 11/7/2013 16:19:00 - 16:41:00

Chemical-looping combustion is a method for the oxidation of hydrocarbons with in-situ $CO_2$ capture. The basic concept of the process involves two interconnected reactors, a Reducer and an Oxidizer, with an oxygen carrier (OC) Â¨C a metal/metal oxide circulating between the two. The main principle in chemical-looping is the oxidation of hydrocarbons with a metal oxide in a $N_2$-free environment. In this study, a transient non-ideal reactor model was developed for two types of Reducers: fixed and fluidized bed [1,2]. The models capture the most important dynamic and spatial phenomena in CLC reduction. The mass balance of 9 species and the energy balance are simulated in both models by the same kinetics mechanisms and reaction constants.

In this work, Ni-based CLC tests in fixed-bed and spouted-bed CLC reactors operated in the University of Connecticut are discussed and the simulation of different fixed and fluidized bed unit data reported in the literature and data of this study are presented. Comparison between experimental and predicted results reveals that the process models and kinetics network proposed are successful in predicting chemical-looping combustion in fixed- and fluidized bed reactors. Through model-assisted comparison between the two types of reactors, some substantial issues in the CLC system performance are discussed. For instance, the comparison of carbon formation (Figure 1) between the two types of reactors (reactor configurations of Iliuta et al. [3]) shows that carbon deposition is not significant in the fluidized bed reactor; whereas, it is of concern at higher conversions in a fixed-bed reactor. Based on the developed reactor models, fixed-bed scale-up calculations are carried out. Spatial and transient pressure and temperature variations are shown and analyzed. Other issues, like challenges in operating CLC at minimum oxygen carrier loading (therefore, smaller Reducer size), OC loading required for high $CH_4$ conversion and OC conversion, time of starting of significant carbon formation at a given OC loading, and $CO_2$ selectivity of fluidized-bed and scale-up fixed-bed units will be discussed.

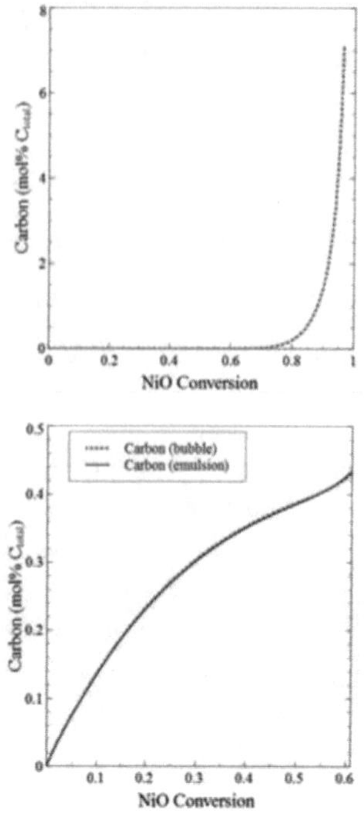


**Figure 1:** *Carbon formation of NiO based OC chemical-looping reduction process in (a) fixed-bed and (b) fluidized bed reactors[3].*

Acknowledgement: This material is based upon work supported by the National Science Foundation under Grant No. 1054718.

[1] D. Kunii, O. Levenspiel, Bubbling bed model: model for the flow of gas through a fluidized bed, I & EC Fundamentals. 7 (1968) 446¨C452.

[2] D. Kunii, O. Levenspiel, Bubbling bed model for kinetic processes in fluidized beds: Gas-Solid Mass and Heat Transfer and Catalytic Reactions, I & EC Process Design and Development. 7 (1968) 481¨C492.

[3] I. Iliuta, R. Tahoces, G.S. Patience, S. Rifflart, F. Luck, Chemical-looping combustion process: kinetics and mathematical modeling, AIChE J. 56 (2010) 1063Â¨C1079.

**(727e) Modeling A Scaled 300 kWth Circulating Fluidized Bed Reactor With Barracuda**
Hamilton, M. A., University of Utah
Lighty, J. S., University of Utah
Originally presented on: 11/7/2013 16:41:00 - 17:02:00

# MODELING A SCALED 300 $kW_{th}$ CIRCULATING FLUIDIZED BED REACTOR WITH BARRACUDA

## CHEMICAL LOOPING COMBUSTION WITH OXYGEN UNCOUPLING

## ABSTRACT

As global warming is continuing to gain recognition as a problem, new technologies for capturing greenhouse gasses such as $CO_2$ are being introduced. The aim of these techniques is to capture $CO_2$ before it enters the atmosphere thus ameliorating its deleterious effects on global temperature and climate. Chemical looping combustion with oxygen uncoupling (CLOU), a subset of chemical looping combustion (CLC), allows for isolation of carbon dioxide ($CO_2$) as an almost pure gaseous stream, suitable for compression and storage. This is done by using two interconnected fluidized beds (CFB) with a metal oxygen carry capable of transferring oxygen from an air reactor to a fuel reactor. The flue gas of the fuel reactor contains a highly concentrated stream of $CO_2$ and $H_2O$. After condensing $H_2O$ the $CO_2$ can be easily captured. This eliminates any direct efficiency loss from the combustion process. The goal of this study was to study fluid dynamics of a 300 $kW_{th}$ CFB using computation particle fluid dynamic software Barracuda VR, then scaling this reactor to a cold flow model and demonstrating with Barracuda that similar fluid dynamics could be obtained.

The scaling was performed similar to Chalmers University of Technologies which was adapted from Levenspiel's text Fluidization Engineering. This method uses dimensionless $\pi$-groups to determine the similarities of the beds. Then from these an overall scaling value is determined and used to adjust the geometric dimensions of the CFB. The time scale dimension is determined from raising the overall scaling value to the ½ power. The particle density and fluid density ratio must be maintained for the cold and hot set up. The hot system has a gas density of 0.236 $kg/m^3$ and a particle density of 2140 $kg/m^3$. From these scaling parameters it was determined that helium (0.167 $kg/m^3$) was the required fluidization gas and that a particle with density of 1450 $kg/m^3$ would be required. The particle of the hot system will fall within the Geldart A group and the particle in the cold system will fall in the Geldart A group. Due to the price and difficulty of obtaining helium it was also determined that the scaled reactor would be run using air at 20 °C as the fluidizing gas. The models to be meshed in Barracuda were created using Solidworks 2012.

Using Barracuda the following transient data was obtained and analyzed: oxygen carrier volume fraction (at points in the air reactor, at points in the air reactor riser, at point in the fuel reactor, and loop seals); the gas pressure at different points of the reactors. The oxygen carrier mass was monitored at the outlet of the air reactor riser and the cyclone outlet. The average residence time of the oxygen carrier was determined in the air reactor and fuel reactor. The reactors were visual inspected using movies created in Barracuda. These were compared for the hot reactor, cold reactor with helium as the fluidizing gas, and the cold reactor with air as the fluidizing gas.

From the Barracuda simulations it was determined that the hot and both cold reactor systems function with similar fluidization patterns. It is recommended that the cold flow model should be constructed to study the effects of static electricity and other unknown inputs. Then kinetics can be added to Barracuda to fully study the hot system.

**(727f) The Ohio State University Chemical Looping Processes: 25 Kwth Sub-Pilot Performance Studies for Natural Gas Conversion and 250 Kwth Pilot Unit Developments**

Kathe, M., The Ohio State University
Tong, A., The Ohio State University
Zeng, L., The Ohio State University
Wang, D., The Ohio State University
Bayham, S., )William G. Lowrie Department of Chemical and Biomolecular Engineering, The Ohio State University, Columbus, OH
McGiveron, O., Ohio State University
Fan, L. S., The Ohio State University
Originally presented on: 11/7/2013 17:02:00 - 17:24:00

The increasing focus on developing alternative fossil fuel conversion technologies is fueled by growing concerns on controlling carbon emissions from traditional technologies in a cost-effective manner. Amongst the newer technologies, chemical looping has been established as a promising candidate exemplified by larger scale demonstrations world-wide. Chemical looping processes utilize metal based oxygen carriers to inherently produce a concentrated $CO_2$ stream coupled with efficient heat-integration which serves to reduce costs and justify potential commercial scale application. The Ohio State University (OSU) has developed multiple product flexible iron-based chemical looping processes for utilizing fossil fuels like coal and natural gas in a unique moving bed reactor design. OSU's chemical looping technologies are driven by the thermodynamic motivation of using higher oxygen carrying capacity of iron-oxide. This enables the process to enlist the flexibility of converting a variety of fuels like coal, natural gas and biomass to generate electricity, hydrogen, syngas or a combination of products with zero or negative $CO_2$ emissions. The iron-oxide based oxygen carriers have been developed to sustain hundreds of redox cycles without loss in reactivity and structural integrity. Amongst the OSU processes, the Syngas Chemical Looping (SCL) process has been developed for converting gaseous fuels to high purity $CO_2$ and demonstrated for more than 350 hours on a sub-pilot (25 $kW_{th}$) scale. With support of the Advanced Research Projects Agency – Energy (ARPA-E) of the United States Department of Energy (USDOE) a 250 $kW_{th}$ high-pressure SCL pilot unit is being constructed at the National Carbon Capture Center. The pilot unit is designed to handle a 250 $kW_{th}$ syngas produced from a commercial Kellogg, Brown & Root transport gasifier to generate high purity hydrogen coupled with a concentrated $CO_2$stream. This will validate the chemistry and operational capability on a commercially applicable scale, thereby completing an important step in the pathway to commercialization.

This presentation will summarize multiple parametric studies which elucidate the interplay of factors like fuel residence time, oxygen-carrier to fuel ratio and temperature for optimizing the natural gas conversion and thereby the process efficiency in the SCL sub-pilot system. The presentation will also include theoretical reactor modeling studies coupled with an overview and comprehensive commercial techno-economic analyses of other OSU chemical looping technologies. Finally, the design development and the progress of the pilot plant construction will be presented.

**(727g) Combustion Of Polyurethane By Chemical Looping Process In A Moving Bed Reactor Using Iron-Based Oxygen Carrier**

Ku, Y., National Taiwan University of Science and Technology
Chiu, P. C., National Taiwan University of Science and Technology
Wu, H. C., National Taiwan University of Science and Technology
Tseng, Y. H., National Taiwan University of Science and Technology
Kuo, Y. L., National Taiwan University of Science and Technology
Originally presented on: 11/7/2013 17:24:00 - 17:45:00

Polyurethane (PU) as a thermal setting plastic with high recovery cost was employed as fuel for chemical looping combustion. The heat value of PU was estimated to be 0.28 kJ/g based on the ultimate analysis of PU by an elemental analyzer. Alumina supported iron-based composite oxygen carriers were prepared for combustion with PU by a moving bed reactor. PU was gasified and reformed by gasification reactants, $CO_2/O_2$, to be CO and $H_2$ as fuels prior to combustion with iron-based composite oxygen carriers at 900°C. By moving bed reactor operation, $CO_2$ yield and PU conversion were both in the moving bed reactor. Long-term operation for PU combustion was lasted for 24 hours, and 100% of $CO_2$ yield was stably maintained. PU combustion with iron-based oxygen carrier in moving bed operation was technically feasible; therefore, chemical looping combustion using solid waste as fuel can be considered as an alternative technology for solid waste treatment.

**(743a) Towards Engineering Advanced Vanadium Flow Battery**

Li, L., UniEnergy Technologies, LLC

Originally presented on: 11/7/2013 15:15:00 - 15:37:00

Flow batteries were specifically invented by electric utilities in the 1960's to store large amounts of electricity in a safe and practical way. The complexities and latent dangers of bulk electricity storage in confined spaces have become more evident in recent years with a series of thermal runaway events at multi-MW battery installations. It is precisely in these areas that the inherent control stability, charge uniformity, thermal consistency and flexible form factor of flow batteries is most compelling.

At small scale, flow batteries may appear complex compared with "a few lead-acid cells in series". However, when extended life and significant scale is required, all batteries rapidly require a complex array of components to actively manage and maintain performance. In the case of flow batteries, these components are invariably simpler, cheaper, more reliable, and easier to control. Consequently, a thoughtful flow-battery system design can yield a significantly more compelling product than is possible with conventional static battery technologies.

This paper discusses system level and cost-effective engineering of a compact, resilient and serviceable bulk electricity storage system. The paper further introduces the UniEnergy Technologies (UET) containerized product.

**(743b) Development and Experimental Validation of An Automotive System Model for Slurry-Based Chemical Hydrogen Storage Materials**

Brooks, K., Pacific Northwest National Laboratory
Pires, R.,
Westman, M.,
Holladay, J.,
Originally presented on: 11/7/2013 15:37:00 - 15:59:00

This paper presents system models for the slurry-based chemical hydrogen storage materials being explored for fuel cell vehicles. As a part of the Hydrogen Storage Engineering Center of Excellence program sponsored by US Department of Energy, PNNL has developed and modeled systems for both exothermic and endothermic storage materials using Simulink® software. The exothermic and endothermic materials modeled were ammonia borane and alane, respectively. These models track the chemical constituents through all the components of the system including the reactor, phase separator, heat exchanger, and product and ballast tanks. To validate the models, experiments with both simulants and actual materials were performed on scaled version of key components. The model was then used with drive cycle simulations for an 80 kWe fuel cell vehicle to determine if the system can meet the DOE targets. Additionally, a sensitivity analysis was performed to provide an operating envelope for other yet-to-be-developed chemical hydrogen storage materials.

**(743c) Deployment of An in-Membrane Micro Fuel Cell for Portable Applications**

Omosebi, A., New Jersey Institute of Technology
Besser, R. S., Stevens Institute of Technology
Originally presented on: 11/7/2013 15:59:00 - 16:21:00

Advances in nano/microfabrication technology has led to the deployment of smaller, faster, and more efficient portable, wireless, and autonomous devices. Today's devices have significantly improved multitasking and computing capability that leads to increased energy requirements. These devices currently rely on secondary batteries like the Ni-Cd, Li-ion, and Li-polymers as a power source, but the need for constant replacement and recharging, interrupted operation, disposal issues, and increasing power requirements have prompted the development of alternative power sources like polymer electrolyte fuel cells as replacements. Fuel cells are high energy power sources that produce DC electricity directly from the stored chemical potential in their fuel and oxidant input. A conventional Nafion-based polymer electrolyte membrane (PEM) fuel cell is a sandwich-like stack of separate layers that facilitate electronic, protonic, and fluid transport. In the conventional micro PEM fuel cell, the auxiliary layers that serve as flow field, current collectors, and reactant distributors contribute significantly to the size and weight of the device leading to bulkier devices. New planar type micro fuels cell architectures have recently emerged where proton conduction is in-plane versus the traditional through-plane orientation. These planar micro fuel cells can facilitate the deployment of thinner and reduced form factor power sources. Distinct from previously fabricated micro fuel cells by other researchers, in this work we present a novel micro fuel cell design where the micro-flow channels for fuel and oxidant input and the membrane electrode assembly (MEA) are fabricated entirely within a Nafion membrane. Nafion is the most widely used membrane in fuel cells that utilize a polymer electrolyte due to its superior properties. The fabrication process leverages electron beam lithography and dry etching for Nafion patterning and this new fuel cell design can result in the use of fewer materials for the deployment of a low-weight, high energy density power source.

**(743d) Experimental Study of Solid Fuel Agglomeration and De-Agglomeration Behaviors in the Reactions With Metal Oxide Particles**

Sun, Z., The Ohio State University
Ding, Y.,
Fan, L. S., The Ohio State University
Originally presented on: 11/7/2013 16:21:00 - 16:43:00

The present work studies the solid fuel agglomeration and de-agglomeration behaviors in the context of chemical looping combustion. A Thermal Gravimetric Analysis (TGA) apparatus is used in this study. PRB coal, Pittsburgh #8 coal, Alabama metallurgic coke, and biomass are used as solid fuels, whereas pure $Fe_2O_3$ and supported $Fe_2O_3$ are used as oxygen carrier particles, respectively. In each test, four types of samples are tested: 1) solid fuel powder only; 2) solid fuel powder + 30 wt% oxygen carrier powder; 3) solid fuel powder + 1-2 mm oxygen carrier balls; 4) solid fuel powder + 30 wt% oxygen carrier powder + 1-2 mm oxygen carrier balls. At reducing environment, each sample is heated to 1000 °C at 50 °C/min, maintained for 10 minutes, and then cooled down to the room temperature. It is found that PRB coal powder and Alabama coke powder form no agglomeration, whereas the biomass powder forms a loose and soft agglomeration only when heated alone. Pittsburgh #8 coal powder forms agglomeration when heated alone or with 1-2 mm oxygen carrier "balls". However, when mixed with the oxygen carrier "powder", no agglomeration forms for Pittsburgh #8 coal powder. To prevent the agglomeration of Pittsburgh #8 coal, two anti-agglomeration approaches are tested: 1) low-temperature devolatization (below caking temperature), and 2) high-temperature devolatization (above caking temperature) and subsequent pulverization. Both approaches are found to be able to effectively prevent/mitigate agglomeration of Pittsburgh #8 coal.

**(743e) Development of Fe-Cu Oxygen Carriers for Solid Fuel Gasification and Hydrogen Production**

Galinsky, N., North Carolina State University
He, F., North Carolina State University
Li, F., North Carolina State University
Originally presented on: 11/7/2013 16:43:00 - 17:05:00

Carbon dioxide emission from fossil energy conversion is a major contributor to the greenhouse effect. Therefore, development of processes to effectively convert fossil fuels with smaller carbon footprints is highly desirable. Chemical looping schemes, which include chemical looping combustion (CLC) and chemical looping gasification (CLG), have shown promise to convert carbonaceous fuels while capturing $CO_2$ efficiently. CLG, in particular, uses a solid metal oxide based oxygen carrier in a redox process to convert carbonaceous fuels into $CO_2$, hydrogen, and heat in separate reactors, avoiding the energy intensive product separation steps.

Currently, Fe based materials are oxygen carriers of choice for CLG processes since reduced iron oxides can effectively convert steam into hydrogen. The limitation of using Fe-based oxygen carriers is that they are relatively slow when converting carbonaceous fuels, especially those in solid forms. In order to address this challenge, we propose to incorporate a secondary metal oxide with oxygen uncoupling properties, e.g. CuO, to overcome the slow reactivity of Fe-based materials. A recent simulation study has shown that the addition of a small amount of CuO to a $Fe_2O_3$ oxygen carrier (9:1 molar ratio between Fe and Cu) can increase the efficiency of the process, resulting in higher $H_2$ production rates.

In the current study, a small amount of CuO based oxygen carrier is incorporated to a batch of Fe based material. TGA and fluidized bed experiments are conducted for chemical looping conversion of coal and methane. Significant increase in reactivity with coal char is observed with 5 w.t. % CuO addition. After the reaction, Fe-Cu spinel structure is not observed under XRD.

**(743f) Synthesis and Redox Pathways of MgAl2O4-Stabilized, Fe-Cu-Based Oxygen Carriers for Chemical Looping Water Splitting**

Imtiaz, Q., ETH Zurich
YÃ¼zbasi, N. S., ETH Zurich
Broda, M., ETH Zurich
Kierzkowska, A., ETH Zurich
Abdala, P.,
van Beek, W.,
MÃ¼ller, C. R., ETH Zurich
Originally presented on: 11/7/2013 17:05:00 - 17:27:00

Hydrogen may become an important energy carrier in the future. The current technologies for the large scale production of $H_2$, *i.e.* steam reforming of natural gas and steam gasification of coal, are very energy intensive owing to the need for various separation steps to obtain pure $H_2$.[1] However, for $H_2$ to become a major energy carrier it has to be produced in an efficient and sustainable way. Chemical looping water-splitting (CLWS) is an emerging technology that can inherently produce a pure stream of $H_2$, thus eliminating the need for energy intensive purification steps.[2] In CLWS, lattice oxygen of an oxygen carrier, typically a transition metal oxide, is used to combust a hydro-carbonaceous fuel. In the subsequent step, the reduced oxygen carrier is re-oxidized with steam to produce a pure stream of $H_2$. Iron oxide is an attractive material for the CLWS process owing to its wide availability, low cost, minimal environmental impact and high equilibrium partial pressure of $H_2$.[3] Nevertheless, its reactivity and cyclic redox stability are two major concerns with regards to the CLWS process. Pure $Fe_2O_3$ that is completely reduced to Fe deactivates after a few cycles, probably due to sintering.[4] To circumvent this rapid deactivation it has been proposed to stabilize iron oxide with a high Tammann temperature support, such as $Al_2O_3$, $ZrO_2$, $TiO_2$, $SiO_2$ or $MgAl_2O_4$. However, it was found that supported $Fe_2O_3$ has often a high tendency for carbon deposition. An approach to improve the reactivity of iron oxide with $CH_4$ and higher hydrocarbons is the addition of more reactive transition metal oxides.[5]

Here, we synthesized $MgAl_2O_4$-stabilized, Fe-Cu-based oxygen carriers containing 70 wt. % $Fe_2O_3$ for CLWS using $CH_4$ as a fuel. The reactivity, $H_2$ yield and carbon deposition characteristics of the oxygen carriers were compared to pure $Fe_2O_3$ and $CuFe_2O_4$ and $MgAl_2O_4$-stabilized $Fe_2O_3$. The oxygen carriers were characterized using powder X-ray diffraction (XRD), $CH_4$ and $H_2$ temperature programmed reduction,

$N_2$ adsorption measurements, scanning electron microscopy, transmission electron microscopy, Raman spectroscopy and X-ray absorption spectroscopy (XAS). The redox and carbon deposition characteristics of the synthesized oxygen carriers were determined at 900 Â°C using both a thermo-gravimetric analyzer and a packed bed reactor.

Pure $CuFe_2O_4$ was found to have a low surface area, compact morphology, slow kinetics and high tolerance towards carbon formation. On the other hand $MgAl_2O_4$-supported $Fe_2O_3$ possessed a 'grainy' surface texture, high surface area, high reactivity towards $CH_4$ and a high activity for carbon deposition. The surface morphology of $MgAl_2O_4$-supported, Fe-Cu-based oxygen carriers were found to be an intermediate of $MgAl_2O_4$-supported $Fe_2O_3$ and $CuFe_2O_4$. Thermo-gravimetric and pack bed measurements revealed that the oxidation kinetics and $H_2$ yield of $MgAl_2O_4$-stabilized, Fe-Cu-based oxygen carriers were, respectively, substantially faster and higher than for $CuFe_2O_4$. For $MgAl_2O_4$-stabilized, Fe-Cu-based oxygen carriers *in-situ* XRD and XAS measurements revealed that $Fe_2O_3$ was stabilized by both a spinel phase ($MgAl_2O_4$) and a delafossite phase ($CuFeO_2$). Additionally, $MgAl_2O_4$-stabilized, Fe-Cu-based oxygen carriers effectively inhibited the decomposition of $CH_4$ and the formation of iron carbide. Scanning electron micrographs and XAS measurements indicated that the reduced quantity of carbon deposition on the bimetallic oxygen carriers is due to an alteration of the surface morphology, impeding the Fe-based activation of $CH_4$.

**References**

[1] A. T-Raissi and D. L. Block, Power and Energy Magazine, IEEE **2004**, 2, 40.

[2] A. Thursfield, A. Murugan, R. Franca and I. S. Metcalfe, Energy Environ. Sci. **2012**, 5, 7421.

[3] R. D. Solunke and G. T. Veser, Ind. Eng. Chem. Res. **2010**, 49, 11037.

[4] C. D. Bohn, C. R. M?ller, J. P. Cleeton, A. N. Hayhurst, J. F. Davidson, S. A. Scott and J. S. Dennis, Ind. Eng. Chem. Res. **2008**, 47, 7623.

[5] S. Wang, G. Wang, F. Jiang, M. Luo and H. Li, Energy Environ. Sci. **2010**, 3, 1353.

**(743g) Study On Performance of the Novel Steam Generation System Using Water-Zeolite Pair**

Nakaso, K., Kyushu University
Nakashima, K., Kyushu University
Tanaka, Y., Kyushu University
Iwama, Y., Kyushu University
Fukai, J., Kyushu University
Originally presented on: 11/7/2013 17:27:00 - 17:49:00

The efforts for energy conservation are requested because of the limitation of fossil fuel and the environmental issue of global warming. The petrochemical and steel industries are especially the largest energy-consuming manufacturing industries. While large amount of waste water with the range of 60-90°C is released from these industrial processes, steam is demanded and generally provided from large quantities of fossil fuel. The system to generate high-temperature steam from waste heat is therefore required for the effective utilization of energy. In this study, a novel steam generation system using a direct heat exchange system with adsorbent-water pair is proposed. Contact between water and adsorbent results in evaporation of excess water due to the release of adsorption heat from adsorbent. Because this system does not require any heat exchangers, not only higher packing density of adsorbent particles in the reactor but also lower heat capacity of the reactor is expected. In our previous study, generation of steam under cyclic operation of steam generation and regeneration processes have been experimentally confirmed using zeolite as adsorbent. Superheated steam at 0.1-0.4MPa was generated using water at 80°C and dry gas at 130°C with lab-scale apparatus [1].

The purpose of this study is to investigate performance of the system using water-zeolite pair. For discussing performance of the system, the mass of the steam generated in the cyclic steam generation and regeneration processes is particularly studied. According to the previous study, the system satisfied with the energy balance in the steam generation process [1]. This implies that mass of steam generated can be estimated based on the energy balance between the initial and final states of the steam generation process. While the final state of the steam generation process is independent from the regeneration process, the initial state of the steam generation process strongly depends on the local temperature distributions and distributions of water content in the zeolite after the regeneration process. Therefore estimation of regeneration process is carried out first. Performance of the system is then discussed based on the estimation of regeneration process.

A mathematical model considering transport phenomena during regeneration process is developed. As the governing equations, conservation equations of mass and energy for gas and solid are considered. Ergun's equation is used instead of the momentum conservation equation for gas. The rate of desorption in the model is determined based on the measurement data from thermal analysis. The rate of drying of zeolite is assumed as constant-rate of drying. For the validation of the model, numerical results are compared to experimental one using the laboratory-scale apparatus. As a result, numerical results reasonably agree with the experimental ones with various gas flow rate and temperatures.

Performance of the system is then discussed in terms of mass of steam generated in the cyclic operation under various regeneration conditions. Because the regeneration process is considered as the energy storage process in the system, mass of steam generated increases with the regeneration time and reaches the saturated value. The typical result is shown in Fig1(a). When the flow rate is 0.51 Nm/s, the mass of steam generated at 140°C (0.36MPa) increases with the regeneration time and reaches the saturated value of 0.13 kg-steam/kg-zeolite. From this result, appropriate regeneration time seems 600-1000 s. To discuss more clearly, the mean steam generation rate is introduced, which is defined as the mass of steam generated during the cycle operation per operation time. For this case, the mean steam generation rate exhibits the maximum value of $0.53 \times 10^{-4}$ kg-steam/(kg-zeolite·s) at the regeneration time of 690 s (see Fig1(b)). Accordingly, the maximum mean steam generation rate is used for the following discussion.

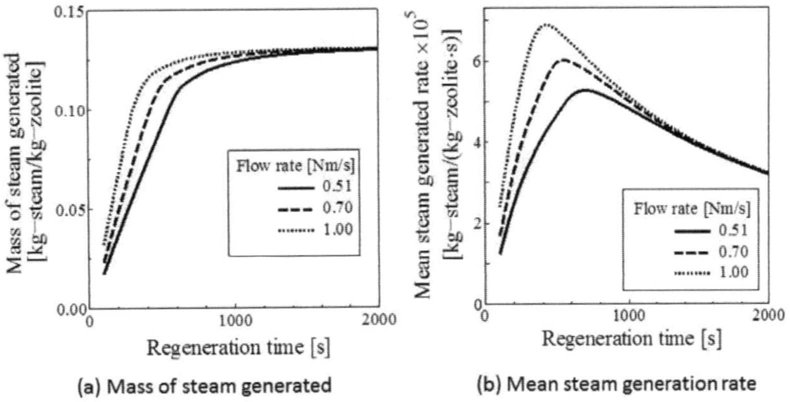


Fig.1 Effect of regeneration time on the mass of steam generated at 140°C in cyclic operation (Inlet gas temperature of 200 °C and Dew point of 60 °C for regeneration)

Effect of the regeneration condition on the maximum mean steam generation rate is studied. The regeneration condition such as temperature and dew point of the gas are studied. The typical result is shown in Fig2. The maximum mean steam generation rates tend to decrease when dew point of inlet gas increases above -20Â°C. When gas temperature is 200Â°C, the effect of dew point on the maximum mean steam generation rate is not so significant. When the gas temperature is 140Â°C, dew point of gas should become low to generate steam at 140Â°C. Consequently, it was found that not only dry gas but also humid gas can be applied to generate steam when gas temperature is enough high. It would be helpful in terms of effective utilization of waste heat such as exhaust gas.

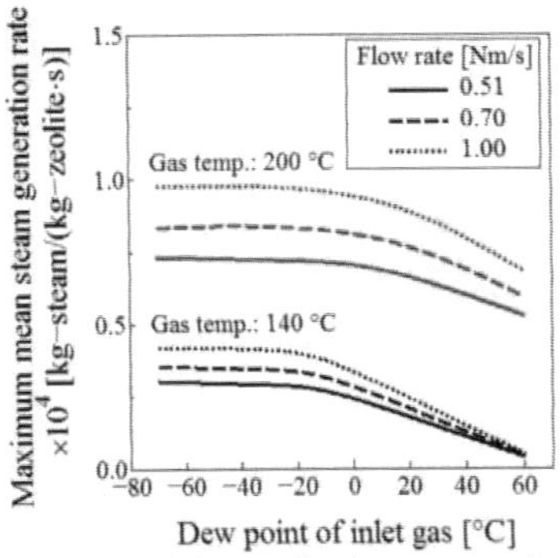


Fig.2 Effect of dew point of the inlet gas for the regeneration process on the maximum mean steam generation rate (Steam temperature: 140°C)

Acknowledgements

This work was partially supported by the NEDO project under Research and Development Program for Innovative Energy Efficiency Technology, and also by the Japan Society for the Promotion of Science (JSPS) Grant-in-Aids for Scientific Research (C) 25420923.

Reference

[1] B. Xue, Y. Iwama, Y. Tanaka, K. Nakashima, A. T. Wijayanta, K. Nakaso, J. Fukai, "Cyclic steam generation from a novel zeolite–water adsorption heat pump using low-grade waste heat", Experimental Thermal and Fluid Science, Vol. 46, 54–63 (2013)

**(774a) A Novel Fluidized Bed Hydrator for Steam Reactivation of Calcium-Based CO2 Sorbent**

Wang, A.,
Wang, D., The Ohio State University
Deshpande, N., The Ohio State University
Phalak, N., The Ohio State University
Wang, W., The Ohio State University
Fan, L. S., The Ohio State University
Originally presented on: 11/8/2013 8:30:00 - 8:55:00

Calcium-based solid sorbents has been widely applied for the removal of acid gases, such as sulfur dioxide ($SO_2$), sulfur trioxide ($SO_3$), and hydrochloric acid (HCl), from coal-fired flue gas. Amid possible carbon dioxide ($CO_2$) regulations, scientists have explored ways to extend this technology for $CO_2$ removal especially from sizable point sources at low cost. Here at the Ohio State University (OSU), researchers have developed a 3-step CaO-based carbonation-calcination reaction (CCR) process that operates in the range of 450 – 950 °C for the simultaneous removal of both $CO_2$ and $SO_2$ from the post-combustion flue gas. The decay in sorbent reactivity for $CO_2$ capture over multiple reaction cycles poses a significant challenge to large-scale cyclic carbonation-calcination process. Thermal sintering at elevated temperatures and the subsequent loss in pore volume both contribute to decreasing $CO_2$ capture capacity. Intermediate hydration has been suggested to be one of the most promising approaches to overcome this limitation by regenerating the particle pore volume and surface area. However, most existing hydration studies have been performed at laboratory scale using thermogravimetric techniques, often investigating conditions not amenable to the 3-step CCR process. In addition, a reactor design specifically suited to high temperature steam hydration is warranted. Experimental data from such a reactor can offer insights to the mass transfer limitations associated with the gas-solid reaction of the extremely cohesive calcium hydroxide ($Ca(OH)_2$) particles under fluidized bed conditions. For the first time, this study reports the design of a bench-scale high-temperature steam hydrator for reactivation of CaO sorbent. The hydrator, consisting of a fluidized-bed reactor with additional internals such as an agitator and a cooling coil, was evaluated using cold-flow tests. Based on the results obtained, a high temperature reactor unit was constructed. The high-temperature steam hydration tests were performed at isothermal conditions in the range of 300-500 °C and 0.6-1.0 atm steam partial pressure. In particular, at an average temperature of 473 °C and $P_{H2O}$ = 1 atm, the 30 min hydration conversion exceeded 70%. Sorbent reactivity toward $CO_2$ was substantially recovered by steam hydration and the correlation follows a linear trend. Lastly, the cooling coil design demonstrated a proof-of-concept for extracting the heat from this highly exothermic high temperature (>450 °C) reaction. This further expands the application of such reactor design beyond $CO_2$ capture, for other applications such as chemical heat pumps.

**(774b) Dynamic Modeling of a Solid-Sorbent CO2 Capture System**

Modekurti, S., West Virginia University
Bhattacharyya, D., West Virginia University
Zitney, S. E., National Energy Technology Laboratory
Miller, D. C., U.S. Department Of Energy, National Energy Technology Laboratory
Originally presented on: 11/8/2013 8:55:00 - 9:20:00

A one-dimensional (1D), two-phase, non-isothermal, dynamic model of a solid sorbent-based $CO_2$ capture system has been developed in Aspen Custom Modeler (ACM). The system consists of a two-stage bubbling fluidized bed (BFB) adsorber-reactor for $CO_2$ capture, a single stage moving bed regenerator for the sorbent regeneration, a multi-stage integral-gear centrifugal compressor system for $CO_2$compression, a pneumatic transport system for solids transport between the adsorber and regenerator, and the balance of the plant. In addition, a heat recovery system has been modeled for the regenerator which includes pre- and post-heat exchangers using steam/boiler feed water as the heating/cooling medium. The solid sorbent is NETL 32D developed at National Energy Technology Laboratory (NETL). The system of partial differential algebraic equations (PDAEs) has been generated using conservation equations, hydrodynamics, and reaction kinetics along with the appropriate boundary and initial conditions.

Flue gas from a supercritical pulverized coal (SCPC) plant enters at the bottom of the lower stage adsorber and regenerated sorbent enters the upper stage adsorber. Embedded heat exchangers have been considered for cooling of the bed using water as the cooling medium. The $CO_2$-rich sorbent is pneumatically transported to the regenerator through a pre-heater. An embedded heat exchanger in the regenerator maintains the unit at near-isothermal conditions. The regenerated sorbent passes through a low pressure steam generator that helps to recover heat for improving the overall efficiency. The medium selected for transporting the solid sorbent between the adsorber and regenerator is the one that minimizes the auxiliary power requirement without violating the overall target for $CO_2$ capture. The $CO_2$ rich gas stream from the regenerator is compressed to pipeline pressure by a multi-stage integral-gear compressor. A glycol absorption system is used to satisfy the pipeline specification for water in the $CO_2$stream. For off-design operation of the compressors, surge detection and control algorithms are developed.

The PDAE system with the appropriate boundary conditions is solved in ACM using the well-known method of lines which involves discretization of the spatial domain and dynamic integration of the time domain. The transient responses of various process variables such as $CO_2$ capture, flue gas and sorbent temperatures, and loading of the sorbent are studied by simulating typical disturbances such as change in the flue gas temperature, flowrate, and composition. A lower level control system is designed to maintain the overall $CO_2$ capture at the desired value in the face of such disturbances.

**(774c) Comparative Environmental Life Cycle Assessment of CO2 Capture Systems Based On Ionic Liquid and Monoethanolamine Solvents**

Farahipour, R.,
Karunanithi, A. T., University of Colorado Denver
Originally presented on: 11/8/2013 9:20:00 - 9:45:00

Carbon capture, storage and utilization (CCUS) is increasingly being viewed as a global climate mitigation option. Central to a viable CCUS system is the development of cost-effective methods for removal of $CO_2$ from flue-gas streams of power plants and other stationary $CO_2$ sources. In addition to economic viability it is important to consider the life cycle impacts of CCUS methods in order to evaluate the net benefits. Moreover, in addition to climate mitigation we would need to consider other issues such as increased life cycle impacts on human and/or ecosystem health. Amin-based (e.g., monoethanolamine) absorption systems for $CO_2$ capture is a fairly mature technology. However, they can be very energy intensive and less attractive from an environmental stand point due to emissions of the used volatile solvent components. In recent years, because of their low volatility, ionic liquids show significant promise as a $CO_2$ absorption solvent for post-combustion $CO_2$ capture. In this work, we present process simulation results of post-combustion $CO_2$ capture with monoethanolamine (MEA) and ionic liquid bmim [PF6]. The heat and material balances derived from these simulations are utilized to construct the life cycle inventory for the capture process. In addition data related to other upstream processes such as solvent (MEA and ionic liquid) production and downstream processes such as $CO_2$ compression, transportation, storage and/or utilization is used to perform a life cycle assessment with an aim to compare the environmental impacts of absorption systems based on these two different solvents.

**(774d) Whole-Chain CCS System Modelling: Enabling Technology to Help Accelerate Commercialisation and Manage Technology Risk**

Matzopoulos, M., Process Systems Enterprise Ltd.
Ramos, A., Process Systems Enterprise, Inc.
Originally presented on: 11/8/2013 9:45:00 - 10:10:00

The commercial implementation of CCS still faces significant challenges. Many of these arise from the fact that the whole chain – and, eventually, whole $CO_2$ transportation network – needs to be considered as a single system in order to make design and operation decisions that satisfactorily address the commercial imperatives and risk requirements of the various stakeholders along the chain. Even a quick analysis shows that design and operating decisions at the power plant can have a significant effect on storage providers at the other end of the chain, and vice versa. A systems modelling approach is essential, but there are currently no tools that can satisfactorily provide this capability over the whole CCS chain.

In order to address this, a £3m project was commissioned and co-funded by the ETI and project participants, who comprise E.ON, EDF, Rolls-Royce, Petrofac (via its subsidiary CO2DeepStore), Process Systems Enterprise (PSE) and E4tech. The project is aimed at delivering a robust, fully integrated tool-kit that can be used by CCS stakeholders across the whole CCS chain that will be released as a commercially-supported software product at the end of the project.

Because of the requirement for many stakeholders to be able to model areas of the chain beyond their own specific processes (for example, a key requirement for power generators is the ability to investigate the effects on their operation of amine and compression systems attached to their plants), the tool-kit will provide process models for the whole chain. This will also allow it to be used by groups who require a whole-chain view, such as engineering companies or government departments who need to quantify policy decisions.

The tool-kit includes models for conventional generation (pulverised coal and combined cycle gas turbine), new generation (gasification and oxyfuel), solvent-based carbon capture, compression, transmission and injection. The individual process models are mostly 'medium-to-high fidelity': multicomponent streams; equilibrium methods for vapour-liquid separation; comprehensive compressor models allowing multi-stage, multi-section compressors with manufacturers' curves; and distributed pipeline models for construction of pipeline networks that can take elevation into account. The capture models use rate-based techniques for accurate quantification of chemical and physical capture in order to quantify energy penalties accurately and allow meaningful analysis of transient operations.

Using the system modelling tool-kit it will be possible to look at single areas such as amine plants in detail; investigate partial-chain operations – for example, power generation, capture and compression; or analyse interactions across the whole chain and eventually the whole CO2 transmission network with multiple sources and multiple sinks.

The tool-kit also caters for interfacing of external software packages such as E.ON's PROATES, in order to simplify interoperability analysis while allowing companies to preserve existing workflows where necessary. In addition an advanced custom modelling capability means that it is easy to add models of new processes and combine them with existing flowsheeting components.

This presentation describes the status of developments to date in whole chain CCS system analysis and plans to roll the tool-kit software out to the wider 'CCS community' following the current testing phase.

**(774e) Optimization of CO2 Capture, Utilization and Sequestration (CCUS) Supply Chain Networks**

Hasan, M. M. F., Princeton University
Boukouvala, F., Princeton University
Floudas, C. A., Princeton University
Originally presented on: 11/8/2013 10:10:00 - 10:35:00

$CO_2$ capture, utilization and storage (CCUS) is an enabling technology toward reducing $CO_2$ emissions from stationary sources which include power plants, cement production plants, iron and steel plants, refineries, petrochemicals and gas processing plants. More than 60% of the total $CO_2$ emissions in the United States are attributed to stationary sources [1]. CCUS connects $CO_2$ sources and utilization/storage sites through three sequential operations: capture, compression, and transportation. $CO_2$ is first captured and compressed at the source plant, and then transported via pipeline to a site where it is utilized or injected for sequestration for geological storage.

It is critical to integrate capture, compression, transportation, utilization and sequestration in an optimal fashion. However, standalone development of CCUS components have made the design of large-scale CCUS networks a complex and challenging task. In this work, we present a mixed-integer optimization (MILP) model for the synthesis of CCUS supply chain networks for nationwide and regional $CO_2$ management. We design the CCUS supply chain network by considering simultaneous selection of source plants, capture technologies, capture materials, and locations of utilization and sequestration sites for the first time. We consider real geographic locations and reliable storage estimates from the available database. Furthermore, our cost model includes the costs of capture, compression, transportation and injection for sequestration, and also revenues from $CO_2$ utilization. The capture and compression costs are derived using advanced modeling, simulation and optimization techniques for a wide range of flue gas compositions and flow rates [2-3], which are crucial to incorporate stationary sources of different types in a large CCUS network. We apply our model to determine the CCUS supply chain network with minimum net expenditure for a given $CO_2$ reduction level.

Our model selects the optimal nationwide CCUS supply chain network from 3317 stationary sources, 491 saline formations, 1837 unmineable coal areas, and 161 utilization locations (oil and gas reservoirs). We also allow four leading capture technologies (absorption, membrane, pressure swing adsorption (PSA) and vacuum swing adsorption (VSA)), and 9 top materials (2 solvents, 4 zeolites [4-5], and 3 membranes) to choose from for $CO_2$ capture. Results suggest that it is possible to capture 50% of the current $CO_2$ emissions from the stationary sources in the United States at a cost less than $30 for each ton of $CO_2$. Results for the CCUS in South Central United States, and Texas are also favorable. We demonstrate that it is crucial to consider both material and technology selections as integral parts of a large-scale CCUS network design.

**References:**

[1]. DOE/NETL. Carbon Sequestration Atlas of the United States and Canada, 2012.
[2]. Hasan, M. M. F.; Baliban, R. C.; Elia, J. A.; Floudas, C. A. modeling, simulation, and optimization of postcombustion $CO_2$ capture for variable feed concentration and flow rate. 1. Chemical absorption and membrane processes. Ind. Eng. Chem. Res. 2012a, 51, 15642-15664.
[3]. Hasan, M. M. F.; Baliban, R. C.; Elia, J. A.; Floudas, C. A. modeling, simulation, and optimization of postcombustion $CO_2$ capture for variable feed concentration and flow rate. 2. Pressure swing adsorption and vacuum swing adsorption processes. Ind. Eng. Chem. Res. 2012b, 51,15665-15682.
[4]. Hasan, M. M. F.; First, E. L.; Floudas, C. A. Cost-effective $CO_2$ capture based on in silico screening of zeolites and process optimization. Submitted for publication, 2013.
[5]. U.S. Provisional Patent Application #61/761,436 and #61/765,284.

**(808a) CO2 Capture By Gas Hydrate Formation: Combined Effect of Thermodynamic and Kinetic Promoters**

Daraboina, N., Denmark Technical University
von Solms, N., Technical University of Denmark (DTU)
Originally presented on: 11/8/2013 12:30:00 - 12:55:00

Carbon capture by gas hydrate crystallization is a promising technology currently being considered in the battle to mitigate climate change. However, the requirement of both high pressure and slow formation rates are serious limitations which need to be resolved prior to implementing this technology at demonstration scale. Various additives (e.g.: THF, TBAB, cyclopentane and others) are known to promote hydrate formation thermodynamically by reducing the hydrate formation pressure. Furthermore there are also kinetic promoters such as surfactants which can increase the hydrate crystallization rate based on the concentration. The appropriate combination of thermodynamic and kinetic promoters has the potential to move the technology towards practical application.. In this work, the hydrate-based process for carbon dioxide capture was tested in a rocking cell (RC-5; Systemtechnik, Germany) apparatus. Various combinations of surfactant (kinetic promoters: SDS and CTAB) and THF (thermodynamic promoter) in liquid water were assessed in order to find the optimum operating conditions.

**(808b) CO2/Brine/Rock Interactions Under CO2 Sequestration Conditions**
Soong, Y., National Energy Technology Laboratory (NETL), Offiice of Research and Development, Department of Energy
Howard, B. H., National Energy Technology Laboratory
Hedges, S. W., U.S. Department of Energy
McIntyre, D., National Energy Technology Laboratory
Warzinski, R. P., U.S. Department of Energy, National Energy Technology Laboratory
Irdi, G.,
Haljasmaa, I.,
Crandall, D., National Energy Technology Laboratory (NETL), Office of Research and Development, Department of Energy
Originally presented on: 11/8/2013 12:55:00 - 13:20:00

## $CO_2$/brine/rock interactions under $CO_2$ sequestration conditions

Y. Soong*, B. H. Howard, S. W. Hedges, D. McIntyre, R. Warzinski, G. Irdi, I. Haljasmaa[1], and D. Crandall[1]

U.S. DOE, National Energy Technology Laboratory,

P.O. Box 10940, Pittsburgh, PA 15236

[1] URS Co.

It has indicated by Intergovernmental Panel on Climate Change Study that injection of $CO_2$ into confined geological formations, given their potentially massive carbon storage capacity and widespread geographic distribution, represents one of the most promising options for mitigating anthropogenically-induced global climate change. The Mount Simon sandstone in the Midwest region of the U.S is recognized as a promising candidate host reservoir for carbon sequestration. The Mount Simon is a deep saline aquifer and is a primary reservoir target for large scale carbon dioxide injection tests due to its proximity to $CO_2$ sources, favorable depth, thickness, permeability, porosity, and the presence of the overlying Eau Claire Formation as a seal. The Midwest Geological Sequestration Consortium (MGSC) has selected the Mount Simon as the reservoir at its major demonstration project. While the Mount Simon Sandstone has ideal reservoir characteristics in some areas, there are significant variations in porosity, permeability, and mineralogy.

Numerous studies have been conducted to investigate changes in host rock properties when exposed to $CO_2$. Luquot and Gouze conducted $CO_2$-enriched fluid flow through a carbonate core sample under in situ sequestration conditions (T = 100 °C and P = 12 MPa). Under these conditions, a decrease in permeability and porosity is seen that is linked to precipitation of Mg-rich calcite. Gouze and Luquot also conducted an X-ray microtomography characterization study of porosity, permeability and reactive surface changes during the dissolution of pure calcite in a brine-$CO_2$ mixture.Â It showed that the increase in permeability is due to the decrease of the tortuosity for homogeneous dissolution, whereas it is due to the combination of tortuosity decrease and hydraulic radius increase for heterogeneous dissolution. Jove Colon et al., studied the permeability, porosity and reactive surface area evolution during dissolution of nonfractured, clay-free Fontainebleau sandstone cores using a flow-through percolation reactor. The experiments were performed at 80°C with 0.1 M NaOH solution. The permeability was found to vary depending on where the dissolution occurred: in the pore void or the pore throat. Liu et al., conducted a simulation study of $CO_2$ sequestration in the Mount Simon sandstone via TOUGHREACT. They indicated that a strongly acidified zone (pH 3-5) forms in the areas affected by the injected $CO_2$ and consequently causes extensive secondary mineral precipitation (calcite, magnesite, ankerite, alunite and anhydrite) and dissolution of feldspars. Thus, the injection of $CO_2$ into a deep saline aquifer affects porosity since it can cause both mineral dissolution and precipitation in the formation. To predict if there will be any changes to the reservoir during sequestration, it is critical to understand the behavior of the authigenic minerals that line pore spaces when they are in contact with the $CO_2$ saturated brine.

An experimental study was conducted to assess the impacts of injected $CO_2$. Core samples from the Mount Simon formation were placed in $CO_2$ saturated brines at sequestration temperature of 85 °C and pressure of 23.8 MPa in the presence of synthetic Illinois Basin brine for a period of six months in a set of 1.3 liter high-pressure vessels (17-4PH-1150 stainless steel, 10.2 cm I.D. x 16.5 cm depth, manufactured by Thar Technologies, Inc.). One sandstone sample with a porosity of 7.9% was obtained at depth of 1,769.7 m from a well located in Vermillion County, IN. The other sample with a porosity of 1.4% was collected at depth of 2634.2 m from a well located in Knox County, IN. The major mineral composition for the first sample is 78% quartz, 15% Feldspar and with less than 2% microcline and trace minerals. The primary mineral composition of the second sample obtained from Knox County is 70% quartz, 22% Feldspar, and around 4% illite/muscovite. Micro CT, CT, XRD, SEM, petrography, porosity, and permeability analyses were conducted prior to the six month exposure. Upon completion of the experiment, the analyses were repeated.

The permeability of the Vermillion sandstone after the six month exposure to $CO_2$ and brine was 0.85 mD compared to 1.6 mD for the fresh sample. On the other hand, the permeability of Knox sandstone was 69 nD compared to 50 nD for the fresh sample. The results suggest that the exposure altered the pore characteristics and thus the permeability of the sample. Micro CT analysis as well as SEM backscattered images were used to probe the Vermillion and Knox Co. sandstone before and after the six month exposure to the $CO_2$/brine environment. It can be seen that mineral dissolution and mineral precipitation occurred after the samples were exposed to $CO_2$/brine for six months. Mineral precipitation occurred in the Vermillion sample pores and cracks blocking flow and resulting within the observed 50% decrease in permeability. For the case of Knox sandstone, the mineral dissolution occurred in pore though the precipitation was observed in the exterior of the sample. This might result in increasing in permeability. In addition, from the analyses of the reacted brines within the reactor, it indicated the possibly of Feldspar dissolution.

When $CO_2$ is injected into a saline formation, acidic conditions form in the brine in the nearby portion of the reservoir thus reducing the pH of the brine in the vicinity of the injection system. Minor minerals in the sandstone, such as K-feldspar, calcite, kaolinite, microcline etc., that are stable in the native environment but not stable at lower pH can dissolve. Eventually, the increase of cation concentration produced by mineral dissolution can supersaturate the fluids with respect to other minerals. If this occurs, precipitation of new minerals can result in a change of porosity and permeability which could affect both $CO_2$ storage capacity and fluid movement. The issue is further complicated by the evolution of pore geometry due to these geochemical reactions that may affect the overall hydrologic properties of the host rock and therefore the performance of the unit as an effective reservoir.

The decreasing permeability found in Vermillion sandstone was probably due to Feldspar dissolution, migration and precipitation altering the sandstone pore/crack structure. For the case of sandstone from Knox county, the permeability increased by 38%. It might be due to the Feldspar dissolution in the pore and mineral precipitation at exterior surface of the sandstone. Investigations are continuing to more clearly elucidate the mechanisms and geochemistry of this system, especially at the pore structure scale.

**(808c) Novel CO2 Capture and Carbonation Process With Waste Concrete**

Shuto, D.,
Yamasaki, A., Seikei University
Hongo, T., Seikei University
Iizuka, A., Tohoku University
Originally presented on: 11/8/2013 13:20:00 - 13:45:00

A novel $CO_2$ capture and carbonation process is developed. In this process, wastes containing calcium, such as waste cement particles or steel slug, are used as a calcium source for carbonation reaction with $CO_2$ emitted from point sources. The process is composed of several steps. $CO_2$ in the flue gas is captured through absorption process with alkaline solution such as sodium hydroxide to form sodium carbonate and sodium bicarbonate. The calcium content in the waste cement particles is extracted with acid such as nitric acid or acetic acid to form solution containing calcium. Mixing of these two solutions leads to the precipitation of calcium carbonate: $CO_2$ fixation via carbonation reaction. Electrodialysis is applied for the regeneration of acid and alkaline from the remained solution after filtration of the precipitant. Thus, the net input of the process is $CO_2$ in the flue gas and waste cement particles, and the output is calcium carbonate. The process feasibility was examined through laboratory -scale experimental studies under various operation conditions. Based on the experimental results, the process was optimized and evaluated in terms of the power consumption and cost for the $CO_2$ recovery, and compared with conventional processes. Net $CO_2$ sequestration can be achieved by the process under certain conditions.

**(808d) Direct Capture and Conversion of $CO_2$ From Flue Gas Via Ex-Situ Mineral Carbonation Using Chemical and Biological Catalysts**

Swanson, E. J., Columbia University
Brady, P. V., Sandia National Laboratories
Park, A. H. A., Columbia University
Originally presented on: 11/8/2013 13:45:00 - 14:10:00

Converting carbon dioxide from flue gas directly into a solid magnesium carbonate presents a compelling option for the mitigation of power plant emissions. The carbonate product is environmentally benign and thermodynamically stable, thus creating a safe and reliable alternative to geological storage. However, given the low partial pressure of carbon dioxide in flue gas and the low reactivity of magnesium bearing minerals, achieving the kinetics required to make such a process feasible is challenging. In addition, magnesium carbonates formed at low temperature tend to incorporate significant waters of hydration, which increases transportation requirements and limits utilization of by-products. Each of these issues was addressed in the design of a multistep process for the conversion of magnesium silicates to carbonates as a carbon capture and storage option. The mineral dissolution reaction is accelerated using a mixture of organic chelating agents to produce a concentrated magnesium solution. This acceleration appears to occur through a stabilization effect on the pores formed as the silicate dissolves. The concentrated magnesium solution produced during the dissolution of the silicate is used to drive the precipitation of carbonates directly from flue gas. Through strict control over the chemistry of a simultaneous absorption and precipitation process, the phase of magnesium carbonate formed can be controlled. In addition, the mass transfer rate of carbon dioxide from flue gas into the precipitating liquor is accelerated using biological catalysis, which is shown to provide significant savings in process equipment size and operating parameters. Finally, the process as a whole is demonstrated to have a negative overall carbon balance through life cycle assessment.

**(808e) A Green Approach to Mitigate CO2 Using Mixed Microbial Culture**

Raghuvanshi, S., Birla Institute of Technology & Science, Pilani
Gupta, S., Birla Institute of Technology & Science, Pilani
Mishra, S.,

Originally presented on: 11/8/2013 14:10:00 - 14:35:00

Carbon dioxide is one of the GHG's (Green House Gases) & responsible for Global warming.. As of July 2011, $CO_2$ in the earth's atmosphere is at a concentration of 392 ppm by volume which is 35% higher as compared to the mid-1800s. Carbon dioxide, which is estimated to contribute 50-60% of the Global warming in the last century, is majorly released during the combustion of fossil fuel. The $CO_2$ concentration in earth atmosphere can be reduced by biological, chemical and technological techniques available through either mitigating or sequestering emissions. In the recent years, biological route has been identified as one of the most viable options available for the mitigation of $CO_2$. There are six carbon fixing pathway's. Out of which, the Calvin cycle (Calvin-Benson-Bassham) is used by photoautotroph's to fix $CO_2$. There are microbes which are ubiquitous in nature and thrive in extreme environment classified as chemoautotrops. They also classified as Extremophiles. The requirement of industries is to develop a economic, efficient and effective system which not only fix $CO_2$ but also capable to withstand in the severe industrial environment. Mixed culture is proved to be a superior candidate, as the consortium of different engineered species develop a symbiotic relationship which helps each other not only to survive in harsh conditions but also curb down $CO_2$. The another possibilities are simultaneous waste water treatment, production of Methane, and bio-hydrogen production. The present work deals in identification of microbes present in sludge of waste water treatment plant having $CO_2$ fixation capabilities. It also gives the insight of the relation which exists between natural selected mixed culture and to understand the capabilities of $CO_2$ fixation. The batch study has been conducted to measure the capabilities of mixed culture to mitigate $CO_2$ for five different V/V % concentration of $CO_2$ 4, 12, 16, 22 and 25. It was observed that $CO_2$ is completed consumed by microorganisms in 4 days. Mitigation rate for different concentration of $CO_2$, bio-mass growth rate and resulting product has been analyzed on per hour basis and per day basis. Mixed culture is found to be superior candidate than photoautotrophic organisms for $CO_2$ mitigation. They show a promising future to develop a low cost, compact size closed system for *in-situ* $CO_2$ mitigation.

# AUTHOR INDEX


Abdala, Paula ....... 246
Abe, Yoshio ....... 170
Abraham, Martin A. ....... 36
Adetule, Oluyemi ....... 118
Adhikari, Shreya ....... 105, 143
Agblevor, Foster ....... 156
Akcay, Asli ....... 124
Akseli, Ahu Aydogan ....... 4
Alam, Md. Zahangir ....... 133
Alayedi, Ibrahim ....... 84
Allen, David T. ....... 11
Andrade, Lidiane Maria ....... 218
Androulaki, Eleni ....... 64
Anid, Nada Marie ....... 16
Anthony, Jennifer L. ....... 62
Aoyama, Toshiyuki ....... 170
Appuhamilage, Nilusha Sudasinghe ....... 182
Arges, Christopher ....... 21
Asa-Awuku, Akua ....... 190
Aston, Victoria J. ....... 166
Atiyeh, Hasan K. ....... 54
Austin, Sharon ....... 11
Bae, Ki-Kwang ....... 233
Bajpai, Rakesh ....... 159
Bakshi, Bhavik R. ....... 5, 104, 147
Balci, Volkan ....... 124
Bandlamudi, Santosh ....... 61, 128
Banholzer, William ....... 18
Banuelos, Jose Leo ....... 65
Bao, Di ....... 90
Bao, Zongbi ....... 195
Bara, Jason E. ....... 58, 76
Bayham, Samuel ....... 239
Beck, Sara ....... 3
Benjamin, Kenneth M. ....... 61, 128
Bennett, Craig A. ....... 138
Bennett, Joshua ....... 60
Besser, Ronald S. ....... 243
Bhattacharyya, Debangsu ....... 251
Bhavsar, Saurabh ....... 165
Bird, Jeffrey ....... 4
Blechner, Jack N. ....... 50
Blinde, Paul ....... 83
Blowers, Paul ....... 84
Bolanos-Reynoso, Eusebio ....... 187
Bollas, George M. ....... 231, 235
Boukouvala, Fani ....... 254
Brady, Patrick V. ....... 261
Brendel, Brittany ....... 184
Brennecke, Joan F. ....... 60, 89
Broda, Marcin ....... 207, 246
Brooks, Kriston ....... 47, 242
Bu, Jie ....... 197
Bu, Quan ....... 137
Budhathoki, Samir ....... 73
Cabezas, Heriberto ....... 174
Cai, Tianxing ....... 85
Caliskan, Erkan ....... 111
Camargo, Gabriel ....... 186
Cano, Natalia Andrea ....... 151
Cavanagh, Eduardo J. ....... 146
Cerniga, Zachary ....... 135
Chaki, Kazutoshi ....... 225
Chan, Kwong-Yu ....... 24
Chan, Sui Yung ....... 40
Chang, E-E ....... 144
Chang, Yuan ....... 80, 148
Chapman, Chris ....... 140
Chappell, Joe ....... 229
Cheali, Peam ....... 102
Chen, Liwen ....... 82
Chen, Si Cong ....... 44
Chen, Yanan ....... 116
Chen, Yanguang ....... 161
Chen, Yi-Hung ....... 144
Chen, Zhengjian ....... 72
Cheng, Weiguo ....... 127
Chiang, Pen-Chi ....... 144
Chiu, Ping-Chin ....... 240
Cho, Seonghyun ....... 15
Cho, Wonchul ....... 233
Choudhury, Hanif A ....... 131
Chowdhury, Muhammad B.I. ....... 209
Chung, Tai-Shung ....... 40, 42, 44, 216
Chupas, Peter ....... 61
Coker, Adebola T. ....... 134
Colombo, Mauricio ....... 132
Colonna, William ....... 224
Cooney, Michael ....... 61, 128
Cortés, Daniel ....... 13
Coutinho, Joao A.P. ....... 69
Crandall, Dustin ....... 257
Crothers, Andrew ....... 77
Cruze, Nathan ....... 147
Cui, Yi ....... 183
Cummings, Peter T. ....... 65
Curtis, Wayne ....... 229
Dai, Sheng ....... 65
Dan, Seungkyu ....... 15
Daraboina, Nagu ....... 256
De Gregorio, Gilbert ....... 122
De Lasa, Hugo ....... 162, 209, 169
De Wit, Jos S. ....... 42
Deng, Shuguang ....... 182
Deshpande, Niranjani ....... 192, 250
DeSilva, Aruni ....... 60, 89
Ding, Yuanpeng ....... 244
Dong, Haifeng ....... 86, 90
Dong, Kun ....... 56
Dougless, Austin ....... 84

# AUTHOR INDEX


Dowding, Matthew ........ 184
Dowell, Niall Mac ........ 126
Dracopoulos, Vassilios ........ 57
Du, Shoucheng ........ 235
Duan, Dongping ........ 145
Dufreche, Stephen ........ 159
Dungen, Barry ........ 182
Durbin, Thomas ........ 190
Dyson, Anna ........ 4
Eason, Tarsha ........ 174
Economou, Ioannis G. ........ 64
Edgar, Thomas F. ........ 19
El-Farra, Nael H. ........ 7
Elia, Josephine A. ........ 172
Evanko, Brian W. ........ 166
Fan, Jiqing ........ 154
Fan, L. S. ........ 192, 239, 244, 250
Farahipour, Reza ........ 252
Feiner, Roland ........ 129
Feng, Guang ........ 65
Fischer, Ari ........ 235
Fisher, Richard P. ........ 3
Florusse, Louw J. ........ 74
Floudas, Christodoulos A. ........ 172, 254
Forero, C. ........ 186
French, W. Todd ........ 119, 134
Froese, Robert E. ........ 154
Fu, Feng-Jiang ........ 44
Fukai, Jun ........ 170, 248
Fukuda, Sho ........ 225
Fulvio, Pasquale F. ........ 65
Gabitto, J.F. ........ 27
Galinsky, Nathan ........ 245
Gallego, Dario ........ 151
Galvita, Vladimir V. ........ 204
Gangadhar, Gayathri ........ 43
Gangadharan, Preeti ........ 82
Gatica, Jorge E. ........ 105, 132, 143
Geng, Zhongfeng ........ 109
George, Anthe ........ 126
Gernaey, Krist V. ........ 102
Gerwe, Brian ........ 84
Ghoshal, Aloke Kumar ........ 210
Giarola, Sara ........ 98
Gibbemeyer, Erin L. ........ 5
Ginosar, Daniel M. ........ 156
Girelli, Ariel ........ 50
Glatz, Charles E ........ 224
Goel, Prem K. ........ 147
Goldstein, Eli A. ........ 234
Gonzalez, Yuri ........ 186
Gonzalez-Garcia, Pedro ........ 187
Grandelli, Heather ........ 59
Green, Chris ........ 224
Grossmann, Ignacio E. ........ 100
Gu, Shuang ........ 26
Guillén-Gosálbez, Gonzalo ........ 13, 150
Guimerá, Roger ........ 13
Gupta, Suresh ........ 43, 262
Gür, Turgut M. ........ 234
Haddleton, David ........ 110
Hale, William ........ 184
Haljasmaa, Igor ........ 257
Hallett, Jason P. ........ 122, 126
Hamilton, Bruce ........ 38
Hamilton, Matthew A. ........ 237
Han, Hongliang ........ 145
Han, Lu ........ 231, 235
Handler, Robert M. ........ 154
Hanes, Rebecca ........ 147
Hasan, M. M. Faruque ........ 254
Hazelbaker, Eric ........ 73
He, Feng ........ 245
He, Yi ........ 219
Hedges, S. W. ........ 257
Hegde, Uday G. ........ 160
Heinzel, John ........ 22
Helmer, Bradley J. ........ 42
Hemmati, Azadeh ........ 197
Hernández, María Rosa ........ 132
Hernandez, Rafael ........ 119, 134
Hicks, Michael ........ 160
Ho, Van Thi Thanh ........ 23
Hoffman, Donald ........ 22
Holbach, Alexander ........ 111
Holladay, Jamie ........ 242
Holmes, William E. ........ 134
Hongo, Teruhisa ........ 260
Horne, W. Jeffrey ........ 76
Horton, Scott R. ........ 138
Hossain, Mohammad M. ........ 162, 169
Hosseini, Davood ........ 202
Houben, Claudia ........ 110
Howard, Bret H. ........ 257
Hsu, Li-Wen ........ 106
Huang, Ying ........ 28, 86
Hutchins, Margot ........ 83
Hwang, Bing Joe ........ 23
Iizuka, Atushi ........ 260
Imtiaz, Qasim ........ 167, 202, 246
Intoci, Nicholas ........ 184
Irdi, Gino ........ 257
Ishizuka, Masanori ........ 41, 55, 176
Ismail, Ahmed E. ........ 77
Ismail, Nur Liyana ........ 122
Itaya, Yoshinori ........ 170
Ito, Kasumi ........ 170
Iwama, Yoshiho ........ 248
Jami, Mohammed Saedi ........ 133
Janho, Michel Kahwaji ........ 132

# AUTHOR INDEX

Jarrell, Kevin ........ 224
Jensen, Kurt ........ 26
Jeong, Seong Uk ........ 233
Jiang, Lan Ying ........ 46
Jiménez-Esteller, Laureano ........ 150
Jin, Jiajia ........ 193
Jindaratsamee, Pinyarat ........ 91
Jing, Dazheng ........ 203
Johnson, J. Karl ........ 67
Johnson, Ryan ........ 229
Jones, Mark E. ........ 18
Judkoff, Ron ........ 1
Jung, Min-suk ........ 21
Jung, Stephanie ........ 136
Kang, Byung Joon ........ 15
Kang, Kyoungsoo ........ 233
Kanjilal, Baishali ........ 184
Kansha, Yasuki ........ 41, 55, 176
Kao, Yu-Lung ........ 106
Kapoor, Kriti ........ 19
Karavalakis, Georgios ........ 190
Karunanithi, Arunprakash T. ........ 78, 79, 252
Kaspar, Robert B. ........ 26
Kathe, Mandar ........ 239
Kawakami, Yoshiaki ........ 170
Kearns, Josh ........ 3
Kesten, Arthur S. ........ 50
Khan, Nymul E. ........ 229
Kierzkowska, Agnieszka ........ 202, 246
Kim, Chang Hee ........ 233
Kim, Jong ........ 19
Kim, Sang Done ........ 233
Kiran, Erdogan ........ 59
Klein, Michael T. ........ 138
Knutsson, Pavleta ........ 164
Kockmann, Norbert ........ 111
Kokossis, Antonis C. ........ 180
Kolah, Aspi K. ........ 108
Kolbe, Massimiliano ........ 177
Komuro, Shogo ........ 91
Kong, Bo ........ 155
Kostetskyy, Pavlo ........ 157
Kotani, Yui ........ 176
Kozinski, Janusz A. ........ 160
Ku, Young ........ 240
Kuo, Yu-Lin ........ 240
Lam, Khee Poh ........ 2
Lammers, Peter ........ 182
Lamsal, Buddhi ........ 224
Lapkin, Alexei ........ 95, 110
Leazer, John ........ 9
Lee, Doyeon ........ 233
Lee, Jong-Min ........ 72
Lei, Hanwu ........ 137
Leion, Henrik ........ 164, 200
Letterio, Michael ........ 26
Lettieri, Dave ........ 83
Lettieri, Paola ........ 140
Lewandowski, Allan ........ 3
Lewis, Randy S. ........ 54
Li, Chunshan ........ 121
Li, Fanxing ........ 161, 245
Li, Jie ........ 172
Li, Liyu ........ 241
Li, Song ........ 65
Li, Wei ........ 193
Li, Wenying ........ 222
Li, Xue ........ 216
Li, Xuehui ........ 35
Li, Yonghui ........ 109
Li, Zhuo ........ 6
Lighty, JoAnn S. ........ 237
Linden, Karl G. ........ 3
Lindsay, Abby ........ 12
Lira, Carl T. ........ 108
Liu, Qing ........ 219
Liu, Wei ........ 33, 53
Liu, Xiaomin ........ 66
Liu, Yuting ........ 63, 115, 116
Llop, Maria ........ 13
Llovell, Felix ........ 69
Long, Jinxing ........ 35
Lou, Helen H. ........ 82
Lu, Bihong ........ 193
Lu, Xingmei ........ 117
Lu, Yongqi ........ 196
Lumley, Nicholas P.G. ........ 120
Lyngfelt, Anders ........ 200, 203
Ma, Hao ........ 35
Ma, Jing ........ 109
Maginn, Edward J. ........ 68, 73
Mahammadunnisa, Shaik ........ 227
Mahoney, Ryan B. ........ 3
Majumder, Subhajit ........ 43
Marcos, Rosa M. ........ 69
Marin, Guy B. ........ 204
Marti, Mustafa E. ........ 224
Martín, Mariano ........ 100
Marumo, Kenji ........ 170
Masanet, Eric ........ 80, 83, 148
Mason, Melissa ........ 159
Massel, Marjorie ........ 89
Materazzi, Massimiliano ........ 140
Mattisson, Tobias ........ 200, 203
Matzopoulos, Mark ........ 253
McCutcheon, Jeffrey R. ........ 50
McGiveron, Omar ........ 239
Mcgranaghan, Eilis ........ 135
McIntyre, Dustin ........ 257
Mehrani, Poupak ........ 158

# AUTHOR INDEX


Mehrkesh, Amirhossein ........ 78
Mejic, Dragan ........ 3
Mele, Fernando Daniel ........ 132
Mendes, Maria Anita ........ 218
Mesta, Adam ........ 84
Mi, Baoxia ........ 44
Mikkola, Jyri-Pekka ........ 34
Miller, David C. ........ 17, 251
Miller, Dennis J. ........ 108
Minnick, David L. ........ 88, 92
Mishra, Somesh ........ 262
Mitchell, Reginald E. ........ 234
Mizuno, Hiroyuki ........ 41
Modekurti, Srinivasarao ........ 251
Mohamed, Shamseldin ........ 169
Mohammad, Javeed ........ 160
Moholkar, Vijay ........ 131
More, Amey ........ 165
Moschovi, Anastasia Maria ........ 57
Moses, John M. ........ 156
Mountraki, Aikaterini D. ........ 180
Mousavi, Bahador ........ 79
Muhich, Christopher L. ........ 166
Müller, Christoph R. ........ 167, 202, 207, 246
Muppaneni, Tapaswy ........ 182
Myers, John A. ........ 229
Najafabadi, Amin Taheri ........ 199
Nakagawa, Tsuguhiko ........ 170, 175, 212
Nakashima, Kazuya ........ 248
Nakaso, Koichi ........ 170, 248
Narvaez, Paulo C. ........ 113
Nasaruddin, Ricca Rahman ........ 133
Nascimento, Cláudio Augusto Oller ........ 218
Newby, Deborah T. ........ 156
Nguyen, Nhan ........ 11
Niazi, Amir Abolhasani ........ 77
Nikolakis, Vladimiros ........ 57
Nirmalakhandan, Nagamany ........ 182
Noshadi, Iman ........ 184
Notoji, Yuu ........ 212
Noureldin, Mahmoud Bahy ........ 48
Nurumbetov, Gabit ........ 110
Nutaifi, Abdulaziz ........ 48
Nybo, Eric ........ 229
Nyman, Andy ........ 224
Nyman, Marianne ........ 4
Oliva, Diego Gabriel ........ 150
Oliveira, Mariana B. ........ 69
Omosebi, Ayokunle ........ 243
Ong, Rui Chin ........ 42, 216
Orgill, James J. ........ 54
Orjuela, Alvaro ........ 113
Ostanek, Jason ........ 22
Oversby, Cori ........ 3
Owaidah, Mana ........ 48
Ozokwelu, Dickson E. ........ 31
Palazoglu, Ahmet ........ 7
Pan, Chun-Jern ........ 23
Pan, Shu-Yuan ........ 144
Park, Ah-Hyung Alissa ........ 261
Park, Chusik ........ 233
Parnas, Richard ........ 184
Parrondo, Janvier ........ 21
Partha, Patra ........ 224
Peek, Ian ........ 22
Pereira, Carla S. ........ 107
Perez, Gonzalo Vilela ........ 164
Pesado, Lidia Cervantes- ........ 187
Peters, Robert W. ........ 6
Peters, Cor J. ........ 74
Peterson, Brian A. ........ 214
Petho, Sandra L. ........ 198
Petkovic, Lucia M. ........ 156
Petrocelli, Frank ........ 138
Phalak, Nihar ........ 192, 250
Pilli, Siva ........ 47
Pires, Rich ........ 242
Pliego-Bravo, Y. S. ........ 187
Poelman, Hilde ........ 204
Porter, Jason M. ........ 120
Powell, Joseph B. ........ 10
Powell, Kody ........ 19
Pucher, Hannes ........ 129
Pucher, Peter ........ 129
Pynn, Michelle ........ 224
Pyrgakis, Konstantinos A. ........ 180
Qahtani, Saleh A. ........ 48
Qi, Zhang ........ 220
Quddus, Mohammad R. ........ 162, 169, 209
Rabideau, Brooks D. ........ 77
Raeissi, Sona ........ 74
Raghuvanshi, Smita ........ 43, 262
Rajangam, Alex ........ 229
Ramakrishnan, T. S. ........ 223
Ramani, Vijay ........ 21
Ramos, Alfredo ........ 253
Ravaghi-Ardebili, Zohreh ........ 177
Razzak, Shaikh ........ 169
Reddy, Harvind Kumar ........ 182
Reddy, P Manoj Kumar ........ 227
Reeves, Stephen A. ........ 105, 143
Reible, Danny D. ........ 39
Ren, Qilong ........ 195
Revellame, Emmanuel D. ........ 134
Reznik, Gabriel ........ 224
Riazi, M. R. ........ 32
Riedthaler, Melissa J. ........ 105, 143
Rincon, Sahra ........ 186
Rodrigues, Alirio E. ........ 107
Rohani, Aida Sharif ........ 158

# AUTHOR INDEX


Romanin, Vince ... 83
Rother, Gernot ... 65
Rugel, Caleb ... 84
Ruiz-Hernández, Antonio ... 13
Ryden, Magnus ... 203
Sai, Zhang ... 220
Saleem, Mahmood ... 114
Sales-Pardo, Marta ... 13
Salvagnini, Wilson Miguel ... 198
Sanaei, Shabnam ... 149
Santaella, Miguel ... 113
Sathitsuksanoh, Noppadon ... 126
Savelski, Mariano J. ... 146, 157
Schaub, Tanner ... 182
Scheffe, Jonathan ... 166
Schlup, John R. ... 62
Schwaiger, Nikolaus ... 129
Schwebel, Georg ... 164
Scurto, Aaron M. ... 88, 92
Selvaratnam, Thinesh ... 182
Seo, Myung Won ... 233
Seo, Samuel ... 89
Shah, Jindal K. ... 73
Shah, Nilay ... 98
Shannon, Matthew S. ... 58, 76
Sharratt, Paul ... 197
Shayegan, Jalal ... 197
Sheehan, John J. ... 37
Shi, Yao ... 219
Shibata, Seishiro ... 175
Shiflett, Mark B. ... 71
Shimizu, Ayako ... 212
Shimono, Kyohei ... 175
Shimoyama, Yusuke ... 91
Shonnard, David R. ... 11, 20, 154
Short, Daniel ... 190
Shuster, Quinton A. ... 132
Shuto, Daiki ... 260
Siebenhofer, Matthäus ... 129
Sierra, Fabio ... 186
Silva, Viviana T. ... 107
Silveira, Carlos Alberto ... 218
Simmons, Blake ... 126
Simoni, Luke D. ... 89
Sin, Gürkan ... 102
Singh, Seema ... 126
Sladkovskiy, Dmitriy A. ... 52
Slater, C. Stewart ... 146, 157
Slivinsky, Maxim ... 213
Smith, Joseph D. ... 214
Somasundaran, Ponisseril ... 224
Soong, Yee ... 257
Srivastava, Ranjan ... 184
Stadtherr, Mark A. ... 89
Stefanakis, Michail E. ... 180
Stuart, Paul R. ... 149
Su, Baogen ... 195
Su, Jincai ... 44
Su, Wei-Nien ... 23
Subrahmanyam, Ch ... 227
Subramanyam, Ramalingam ... 159
Summers, R. Scott ... 3
Sun, Jian ... 127
Sun, Ning ... 126
Sun, Shi-Peng ... 40
Sun, Zhenchao ... 244
Sundqvist, Sebastian ... 200
Sunol, Aydin K ... 135
Swanson, Edward J. ... 261
Tahir, Muhammad suleman ... 114
Tan, Chung-Sung ... 144
Tanaka, Yuuki ... 248
Tang, Juming ... 137
Tanino, Masayuki ... 170
Tao, Ling ... 185
Taqueda, Maria Elena S. ... 198
Taylor, Richard ... 140
Thibault, Jules ... 158
Thwaites, Fred ... 148
Tillie, Charles F. ... 105, 143
Tomlinson, Sean ... 62
Tong, Andrew ... 239
Tran, Kim ... 126
Tseng, Yao-Hsuan ... 240
Tsouris, C. ... 27
Tsutsumi, Atsushi ... 41, 55, 176
Uddin, Mohammad Shahab ... 133
Upadhyaya, Kamal Lamichhane ... 119
Urban, Robert A. ... 104
Uzun, Alper ... 124
Van Beek, Wouter ... 246
Van Duin, Adri C.T. ... 67
Van Noyen, Jasper ... 203
Vance, Leisha ... 174
Vasenkov, Sergey ... 73
Vega, Lourdes F. ... 69
Velásquez, Hector Ivan ... 151
Vergadou, Niki ... 64
Verma, Sandeep ... 223
Veser, Götz ... 165
Vigil, R. Dennis ... 155
Vilaseca, Oriol ... 69
Vollen, Jason ... 4
Von Solms, Nicolas ... 256
Voorhies, Wayne ... 182
Vu, Diep ... 190
Wakatsuki, Toshiya ... 225
Walker, Michael E. ... 83
Wang, Alan ... 250
Wang, Dawei ... 239, 250

# AUTHOR INDEX


Wang, Han....73
Wang, Jinquan....127
Wang, Lu....137
Wang, Qian....93, 117, 222
Wang, Wei-Cheng....185
Wang, William....192, 250
Wang, Xiangqian....193
Wang, Xiaonan....7
Wang, Yi Yang....84
Ward, Barbara J.....3
Ward, Jeffrey D.....106
Warzinski, Robert P.....257
Weatherley, Lawrence R.....88
Weimer, Alan W.....3, 166
Welton, Tom....126
Westman, Matt....242
Wiesner, Theodore....215
Wu, Hsuan-Chih....240
Xing, Huabin....195
Xing, Yanjun....63, 115, 116
Xu, Jian....123
Xu, Qiang....85
Xu, Rong....215
Xu, Yajing....28
Xue, Tong....72
Yacob, Tesfayohanes W.....3
Yamasaki, Akihiro....260
Yan, Yan....195
Yan, Yushan....26
Yang, Chunzhen....24
Yang, Qiwei....195
Yang, Yiwen....195
Yao, Yuan....80, 148
Yaseneva, Polina....95
Ydstie, B. Erik....2
Ye, Qing....196
Yokozeki, Akimichi....71
Yoon, En Sup....15
You, Fengqi....213
Yu, Yingzhe....109
Yue, Dajun....213
Yüzbasi, Nur Sena....167, 246
Zappi, Mark E.....159
Zeng, Liang....239
Zhang, Bo....67
Zhang, Haibin....224
Zhang, Junping....127
Zhang, Minhua....109
Zhang, Suojiang... 28, 56, 66, 86, 90, 117, 121, 127, 217
Zhang, Xiangping....28, 86, 90
Zhang, Xiaochun....66
Zhang, Xin....86, 90
Zhang, Yong....68
Zhang, Yu....138
Zhang, Zhiqiang....109
Zhao, Jie....2
Zheng, Shudong....219
Zhong, Li....145
Zhou, E....145
Zhou, Zhiquan....231, 235
Zhu, Jia Ming....46
Zitney, Stephen E.....251
Ziv, Guy....104